Marine and Fisheries Policies in Latin America

This book reviews the frameworks and implementation of marine, fishery and coastal laws and policies in Chile, Mexico and Peru.

Chile, Mexico and Peru share biodiverse coastal and marine environments which are being affected by unregulated and informal developments, and thus share similar challenges. Each country is currently at a different stage of advancement in their institutional response to these complex challenges. By providing a comparison of the frameworks, approaches and overall implementation of policies and laws, this book acts as a tool to influence and inform further efforts in conservation and sustainable use of marine resources, particularly fisheries, in these countries and others in Latin America and the Caribbean. A broad range of issues are covered including food security, tourism, fisheries, oil and mineral extraction from the seabed, wind power, coastal and marine pollution and endangered species conservation. The chapters compare how each country addresses these issues from an institutional, legal and policy perspective. The book concludes by identifying common lessons, reoccurring challenges and develops scalable recommendations applicable to the case study countries and the wider region.

The book will be of interest to advanced students, policy makers and researchers in marine and fishery science, law and policy.

Manuel Ruiz Muller is Senior Advisor and Researcher for the Peruvian Society for Environmental Law, Peru.

Rodrigo Oyanedel is a PhD researcher at the Interdisciplinary Centre for Conservation Science, University of Oxford.

Bruno Monteferri is Director of the Marine Governance Initiative and Director of We Conserve by Nature at the Peruvian Society for Environmental Law, Peru.

Earthscan Oceans

For further details please visit the series page on the Routledge website:
www.routledge.com/books/series/ECOCE

Marine and Fisheries Policies in Latin America

A Comparison of Selected Countries

Edited by Manuel Ruiz Muller, Rodrigo Oyanedel and Bruno Monteferri

Routledge
Taylor & Francis Group

LONDON AND NEW YORK

earthscan
from Routledge

SPDA

First published 2020
by Routledge
2 Park Square, Milton Park, Abingdon, Oxon OX14 4RN

and by Routledge
52 Vanderbilt Avenue, New York, NY 10017

Routledge is an imprint of the Taylor & Francis Group, an informa business

First issued in paperback 2021

British Library Cataloguing-in-Publication Data
A catalogue record for this book is available from the British Library

Library of Congress Cataloging-in-Publication Data
A catalog record has been requested for this book

ISBN: 978-1-138-38692-1 (hbk)
ISBN: 978-1-03-208866-2 (pbk)
ISBN: 978-0-429-42652-0 (ebk)

Typeset in Bembo
by Wearset Ltd, Boldon, Tyne and Wear

Contents

Contributors

Alejandro Correa is a lawyer with a Masters' Degree in Environmental Policy, Governance and Markets from the University of Melbourne. Mr. Correa is Co-Founder and Director of Public Policies of Costa Humboldt.

Alfredo Gálvez is a lawyer graduated from the National University of San Marcos and specializes in biodiversity and natural protected areas law. He has a Diploma in Management of Natural Resources from the Autonomous University of Yucatán, Mexico, a Diploma in Law and Hydrocarbons from the National Society of Mining, Petroleum and Energy, Peru, and a Masters' Degree in Natural Protected Spaces from the Autonomous Universities of Madrid, Complutense of Madrid and Alcalá, Spain. Mr. Gálvez works for the Biodiversity and Indigenous Peoples Program of SPDA.

Andrea Cuba is a lawyer from the University of Lima, Peru. She has a Diploma in Regulation and Environmental Management from the University of Applied Sciences (Lima) and a Masters' Degree in Environmental Law and Policy from the University of London. At present, she is a consultant in Environmental Law.

Bruno Monteferri is an environmental law specialist. He graduated from the Pontific Catholic University of Peru. He has a Diploma on Integrated Management of Coastal Zones from the University of Guadalajara, Mexico, a Diploma on Environmental Law and Natural Resources from the Pontific Catholic University of and a Masters' Degree (MPhil) in Conservation Leadership from University of Cambridge. Mr. Monteferri has worked at the Peruvian Society for Environmental Law since 2005 and leads the "We Conserve for/by Nature" and Marine Governance initiatives.

Carlos I. Vargas is an engineer in aquaculture and has a Masters' Degree in Aquaculture and Biotechnology from the University of Chile. At present he works as an independent consultant.

Christel Scheske has a Masters' Degree in Conservation Science from Imperial College, London and in Social Psychology and Development from the

University of Cambridge. She is German-Indonesian living in Peru and works at the Peruvian Society for Environmental Law in the "We Conserve for/by Nature" Initiative and the Initiative for Marine Governance.

Fernando A. Rosete Vergés is a biologist, graduated from Autonomous Metropolitan University-Xochimilco, Mexico. He has a Masters' Degree in Conservation and Natural Resource Management from Michoacan University of San Nicolás of Hidalgo and a Doctorate in Geography from the National Autonomous University of Mexico. Since 1998 he has worked on marine and coastal area management issues. Mr. Rosete Vergés is head professor at the National School of Superior Studies (Morelia Unit) of the National Autonomous where he coordinates the Territorial Management and Planning Unit.

Irene Hofmeijer is a Bachelor in Environmental Sciences from McGill University, Canada and has a Masters' Degree in Environmental Management from Duke University, United States. She is the founder and director of a social enterprise Life Out of Plastic – L.O.O.P, where she currently works.

José Manuel Troncoso is a geographer from the Pontific Catholic University of Chile and at present undertakes research at the Costa Humboldt organization.

Juan Carlos Castilla is a marine biologist with a Doctorate from University College of North Wales, United Kingdom. He is Professor Emeritus at the Pontific Catholic University of Chile. In 2010, Mr. Castilla received the National Prize for Applied Sciences and Technologies of Chile; in 2011, the Ramón Margalef Prize in Ecology; and in 2012, the Mexican Prize for Science and Technology.

Liliana Pardo López is a graduate in biology from Veracruzana University, Mexico, and has a Masters' Degree in Biochemistry from the National Autonomous University of Mexico, a Doctorate in Biochemistry from the same university and a postdoctoral from the Atomic Energy Commission, Paris, France. She is presently Head Researcher "B" at the Biotechnology Institute of The National Autonomous University of Mexico.

Luciano Hiriart-Bertrand is a marine biologist with a Masters' Degree in Marine Biodiversity and Conservation from Scripps Institution of Oceanography of the University of California, San Diego. He is founder and Executive Director of Costa Humboldt.

Manuel Ruiz Muller is a lawyer, Senior Advisor and Researcher at the Peruvian Society for Environmental Law. He has 25 years of experience with regards to promoting rights and environmental policy, having collaborated with IDB, FAO, WIPO, IUCN, UNDP and UNEP, among other organizations. He is at present a Darwin Fellow for Botanical Garden Conservation International.

Maria Fernanda Onofre Villalba is a student of Environmental Sciences at the National School of Superior Studies (Morelia Unit) of the National Autonomous University of Mexico.

Mariana Torres García is a student of Environmental Sciences at the National School of Superior Studies (Morelia Unit) of the National Autonomous University of Mexico.

Pedro Solano is an environmental law specialist who studied at the Pontific Catholic University of Peru. He has been a member of the Peruvian Society for Environmental Law since 1989 and is currently its Executive Director. Mr. Solano is one of the most recognized national and international experts on natural protected areas and is a member of the Commission for Protected Areas and Environmental Law of the World Conservation Union. He is also a board member of the Interamerican Association for Environmental Defense and the Amazon Conservation Association.

Rocío López de la Lama is a biologist at the University Cayetano Heredia from Perú. She has a Masters' Degree in Conservation Leadership from the University of Cambridge, England. She is currently a student of a doctoral degree at the University of British Columbia and associated to the Institute for Resources, Environment and Sustainability.

Rodrigo A. Estévez is a sociologist of the University of Concepción (Chile) and has a Masters' Degree in Biological Sciences from the University of La Serena (Chile) and a Doctorate in Philosophy from the University of Melbourne. He is currently a researcher associated to the Center of Applied Ecology and Sustainability of the Pontific Catholic University of Chile.

Rodrigo Oyanedel is a marine biologist, with a Masters' Degree in Environmental Science and Management from the University of California, Santa Barbara. He is currently pursuing a Doctorate in the Zoology Department at Oxford University (England).

Santiago de la Puente is a biologist from the University Cayetano Heredia in Peru and specializes in fisheries resource management. He holds a Masters' Degree in Management of Natural Resources and Environmental Sustainability from the University of British Columbia (Canada). Mr. de la Puente is currently pursuing a Doctorate in Oceans and Fisheries at the University of British Columbia and is part of the Global Ocean Modeling Group and the Institute for the Oceans and Fisheries.

Stefan Gelcich is a marine biologist with a Masters' Degree in Environment and Development from the University of Cambridge (England) and a Doctorate from the University of Wales, Bangor. He is a professor at the Pontific Catholic University of Chile and in 2015 received a Pew Marine Conservation Fellowship. Mr. Gelcich is also a Board Member of Global Green Grants.

Ximena Vélez-Zuazo is a biologist from the National Agrarian University La Molina (Peru) with a Masters' Degree in Tropical Biology and a Doctorate in Biology from Universidad de Puerto Rico. She is the Marine Director of the Biodiversity Monitoring and Assessment Program–Peru and Associate Researcher for the Smithsonian Institute Center for Conservation and Sustainability.

Foreword

Despite the global social and environmental importance of oceans and fisheries, reflection about these issues is limited and difficult to access particularly in Latin America. The absence of quality policy, legal and institutional literature on fisheries and coastal and marine issues in Latin America is notorious.

A common phrase in many of our countries is that "we live with our backs to the sea", in other words, without adequate social, economic, environmental and affective consideration about the importance of coastal and marine ecosystems and the resources and services they provide. Nor do we think in terms of the urgency of conservation and sustainable use of these ecosystems and resources.

Nevertheless, society is gradually changing its perception about the coastline and the oceans. It is more and more frequent that we see civil society groups promoting campaigns and different conservation projects. State-led initiatives to monitor and protect fisheries and improve wastewater disposal systems which end up in the sea are on the rise. Likewise, small but significant changes are evident in the private sector, where concerns and interest in improving extractive activities and better managing ecosystems are also part of this positive trend.

But coastal and marine zone are much more than fisheries only. Ecosystem services such as climate regulation, free navigation and transport, millennial cultures associated to the coast and marine natural protected areas are just as important as the ocean itself. Over the past few decades, international and national public policies have appeared, with guidance and principles geared towards coastal and marine conservation and sustainability. Their goal is to create safeguards to protect the public interest and the common good expressed in the oceans and seas and their services and resources.

It is in this context that the Peruvian Society for Environmental Law (SPDA), as a part of our efforts to contribute to informed national and regional dialogue, presents this publication with a series of chapters from well-known authors, addressing multiple issues which in some way unite countries with extensive coastlines and seas and millenary traditions pertaining to the use and exploitation of marine resources. Chile, Mexico and Peru have different realities, but also certain common policy and institutional approaches regarding their coastal and marine territories. At the same time, they face similar challenges, including overexploitation of their fisheries, marine pollution, weak enforcement and control mechanisms, social conflicts, etc. This book provides a wide range of views regarding these and other issues.

A growing international concern and focus on coastal and marine issues is reflected in the Blue Economy, the United Nations Convention on the Law of the Seas, the review of the Aichi Biodiversity Targets and the Sustainable Development Goals, among others. They provide a framework from which to promote different initiatives and projects to contribute to coastal and marine conservation and change the paradigm towards one of "facing the sea". If we could all look at our planet from space, we would clearly see that the name we gave it was incorrect, far from being planet "earth", it should have been named planet "water" or "sea". This minor historic detail in itself is essential to help understand the need to research and better manage our oceans, their services and resources.

The future of planet "water" will depend on how well or badly we manage our oceans. We hope this book will contribute to this effort.

Pedro Solano
Executive Director,
Peruvian Society for Environmental Law

Prologue

Fisheries management, food security and sustainability are three key issues that must be addressed during this century. Particularly after the Great Acceleration and post Second World War period, technological advances for prospecting and detection of shoals of fish and massive extraction have led to significant problems of overexploitation of fisheries. According to several analyses, 75 per cent of fish and shellfish stocks around the world are being fully exploited or overexploited. In addition, static and deterministic production models focused on single species, led to the development of the concept of Maximum Sustainable Yield, which has demonstrated its limits and, often, its failures (Defeo *et al.*, 2007).

As I have already argued, within the global concept of "fisheries" there is more than one "fishing-bag". Rather, there are several categories. High seas fisheries (200 nautical miles off the coast), industrial fisheries (within 200 nautical miles) and coastal artisanal and subsistence fisheries must be, at the very least, analyzed by categories. For each one of these categories lies the possibility that the "tragedy of common use resources", originally formulated by Lloyd (1883) and later popularized by Hardin (1968) is expressed. This is exacerbated in the high seas where international fisheries regulation is non-existent. In contrast, countries are clearly responsible for policies, regulations, control, inspections and sustainability in the case of industrial and artisanal fisheries.

Fishing countries in Latin America are not the exception in terms of the deterioration of their fisheries. However, in my opinion, countries like Chile, Mexico and Peru, have shown leadership through the adoption of national strategies, implementation of marine spatial zoning, enforcement actions and novel and innovative regulations for fisheries. Such is the case of co-management schemes and the creation of massive coastal and marine areas for exclusive community access (mainly in Chile) for small-scale artisanal fisheries. Nevertheless, reverting overexploitation of their main fisheries and resources remains an elusive task.

It is clear these countries have different fisheries frameworks but particularly in the case of artisanal fisheries, they show unequivocal commitment to implementing socio-ecological approaches. These are not centered or focused on specific resources, but rather, consider the environment, zoning, exclusive and preferential access to communities, and fishers and communities themselves as key dimensions in management and sustainability possibilities.

This book fills a gap of information in Latin America, with a comparative approach to both industrial and artisanal fisheries in Chile, Peru and Mexico. A group of renowned experts from various disciplines – lawyers, biologists, engineers, geographers and environmentalists – address a series of issues pertaining to legislation, practical implementation, sustainability, illegal fishing, planning, management, surf break protection and conflicts, in an effort to contribute to solving coastal and marine and fisheries problems in the region. Though each country has its own set of problems, different legal frameworks, methods and idiosyncrasies, common threads and approaches are identified.

If there were the need to highlight a common issue affecting fisheries in Chile, Peru and Mexico, I would argue it would be illegal fishing. This is a major problem both in these countries and around the world. Without databases on reliable landings, it is simply impossible to know what goes on in the dynamics of a fishery, even if sophisticated models are used or the miraculous philosopher's stone of "ecosystem management" is invoked. It is one of the greatest challenges for the health of our oceans. There is no easy solution unless some form of strict control is applied, incentives are developed, true involvement of fishers takes place, including artisanal or subsistence fishers and cutting-edge technology is used. In the particular case of artisanal fishers, an additional problem which extends to them is that of governance. Latin America here shows substantial advances, but the road is full of difficulties. World markets and competition between aquaculture and natural (wild) fisheries, is not minor (Castilla *et al.*, 2015).

Ethical decisions by consumers and users are also center stage in the conversation of current and future fisheries management and conservation. The development of marine environmental ethics requires knowing the sea and ocean and falling in love with them. But this is not a regular occurrence, given we are blind to the sea; not only do we turn our backs on it, but even when we do look, we are only scratching its surface. We are not aware of the wonders, benefits and ecosystems the sea offers – including fisheries. Humans are terrestrial animals and therefore often estranged from aquatic environments, which makes the romance with the seas and oceans ever more difficult.

In Chile, I illustrated this with an extremely simple cultural concept: "Chile is Sea". Chile is not long and narrow…. Chile is a long, wide, blue and deep country… Chile is 75 per cent sea (Castilla, 2014). This should be rooted as part of a basic national marine culture. But it is not. I suspect, without having a quantitative survey at hand, that something very similar is happening in Latin America. Marine education should play a critical role in raising awareness – among our children first – about a responsible marine culture and complicity which lead to a love for the sea. This involves generations and cannot be fulfilled overnight.

This book and its dissemination is a contribution for Latin America in different directions: on comparative learnings about fisheries management;

lessons and challenges on common fisheries; engagement of people to actively participate in the sustainability of marine resources; ensuring good decisions when acquiring fisheries products. But fundamentally, bringing to our attention the urgent need for developing solid and permanent marine education and awareness programs in Latin America. It can be done.

Juan Carlos Castilla, Ph.D. D. Sc.
Catholic University of Chile
Santiago de Chile, March 2019

References

Castilla, J.C. (2014), Chile es Mar. In *Mar de Chile*. Banco Santander y Museo Chileno de Arte Precolombino (Eds). Santiago de Chile. pp. 13–42.

Castilla, J.C., Espinoza, J., Yamashiro, C., Melo, O., and Gelcich, S. (2015), Telecoupling Between Catch, Farming, and International Trade for the Gastropods *Concholepas concholepas* (Loco) and *Haliotis* spp. (Abalone). *Journal of Shellfish Research* 35:499–506.

Defeo, O., McClanahan, T., and Castilla, J.C. (2007), A Brief History of Fisheries Management with Emphasis on Societal Participatory Roles. *Fisheries Management: Progress towards sustainability*. T. McClanahan & J. C. Castilla (Eds). Blackwell Publishing, UK. pp. 3–21.

Hardin, G. (1968), The Tragedy of the Commons. *Science* 162:1243–1248.

Introduction

Manuel Ruiz Muller, Rodrigo Oyanedel and
Bruno Monteferri

Despite their social, ecological and economic importance, we have historically "turned our backs to the sea". It is only recently that oceans and seas – which occupy 71 percent of the planet – have started to occupy more visible spaces in global and national conservation and sustainable development programs and strategies. Key milestones in this recognition include the Aichi Biodiversity Targets (2010) and the adoption of the World Oceans Day (June 8) in 2009 by the United Nations. However, efforts by philanthropic organizations and NGOs in the implementation of specific programs to finance and contribute to conservation and sustainable use of oceans and their resources in general are relatively new and still insufficient.

For Latin America and Chile, Mexico and Peru in particular, the oceans, seas and fisheries play a key role in their development both directly and indirectly. For instance, although fishing contributes with relatively low percentages to national Gross Domestic Product (i.e. 0.34 percent Peru, 0.4 percent Chile, 0.2 percent Mexico), it directly and indirectly adds to the generation of employment. Fishing activities – in general terms – generate 350,000 direct jobs and 2,000,000 indirect jobs in Mexico; 300,000 direct and 1,500,000 indirect jobs in Peru; while in Chile 90,000 workers are employed directly and 95,000 indirectly. In addition to employment, fishing activities are essential for food security both locally and internationally, when considering that these three countries are world fishing powers.

On the other hand, the coast and adjacent seas sustain a growing tourism sector which, with variations among countries, has become essential for the development of local coastal economies. Either through large tourist spots (i.e. Cancun or Acapulco in Mexico), specialized tourism (i.e. gastronomic, sports) or cities (i.e. Viña del Mar in Chile or Mancora in Peru) and small villages that offer seasonal alternatives, the communities and local populations who benefit from job opportunities generated from the seas and its services are countless.

Nevertheless, the multiple ecosystem services and benefits provided by the seas, oceans, coastal areas and fishing, have all but been ignored by greater portions of urban societies, even if located on coastal areas. Food, energy, diversion, transport, roads, potential products for the biotechnological and food industry, climate control, among others, are only some of the dimensions where these spaces offer little–understood and untapped possibilities in Chile, Mexico and Peru.

The risks seas, oceans, coastal areas and fishing confront vary between countries, but present common trigger factors. For example, overfishing and the uncontrolled extraction of all types of marine resources, has resulted in almost complete depletion in certain areas. In Peru, fine flounder (*Paralichthys adspersus*) and corvina (*Cilus gilberti*) have reduced their stocks significantly as a result of indiscriminate fishing, in this case artisanal and recreational. In Chile, the blotched stingray (*Urotrygon chilensis*) and pink cusk-eel (*Genypterus blacodes*) also suffer from overexploitation. In Mexico, overfishing and illegal fishing have put all commercially valuable fisheries in the Gulf of Mexico on the brink of collapse. There are endless cases and examples around the world as well.

In the case of Peru and Chile and despite strict regulations, the pelagic species industrial fishery has turned these countries into world superpowers in terms of fishmeal production and canned goods. However, this also puts at risk both the species and the food chain of which they are a part. Even though their stocks have continuously recovered over time and are classified as "healthy", they are being subjected to constant pressures which threaten their long-term sustainability. Climate change and related sea–current variations and temperatures present an uncertain challenge and complex scenario for this critically important fishery. Bycatch especially by industrial vessels, also generates a significant impact on populations of coastal-marine species, including dolphins, sea lions and tortoises. Closures do not seem to have had the dissuasive effect required to guarantee the population viability of different species.

The adjacent seas and coastal-marine ecosystems also suffer growing pressure from an infrastructure development ("bricks and mortar") that modifies the shores, changes the bathymetry, alters sand and wind flows, impacts natural infrastructure, among others. Marinas, piers and beach resorts generally have a negative impact on these spaces, generating adverse externalities that are not planned, accounted for or much less internalized by those who undertake these projects. There are also endless examples. For instance, the hotel infrastructure in Quintana Roo in Mexico has definitely altered coastal-marine ecosystems adjacent to the shores. Beginning in Lima, Peru towards the south, for 150 km, the coastal boarder zone shows an urban growth during the last 20 years, marked by illegality, disorder and disregard for any environmental or natural patrimony protection standard or precaution. Such developments create direct impacts due to infrastructure itself and waste discharge – both legal and illegal – with ultimate pollution consequences for these environments.

There is no tradition of legal protection for coastal-marine spaces and ecosystems in Latin America, although this has recently been changing, for example, through different forms of marine protected areas. For decades, countries in the region have lived without a genuine strategy on how to sustainably develop and benefit from the coasts, oceans and seas – beyond the usual commercial or industrial exploitation. At present, coastal-marine protected spaces have increased significantly in some countries. For example, Chile has 42 percent of its coastal-marine territory under some category of protection; in Mexico nearly 23 percent of its coastal-marine territory is protected, while in Peru the percentage does not even reach 0.5 percent, although with the creation of the Tropical Reserve National Sea of Grau this will hopefully increase to a still meagre 1 percent. These tendencies aim to align countries with Aichi Target 11 which proposes that by 2020, 10 percent of marine ecosystems are under some type of legal protection, including through natural protected areas.

In spite of this somewhat bleak picture and challenging situation, Chile, Mexico and Peru have developed over the past decade and more, a series of instruments and regulations that seek to benefit from these resources, seas and coastal-marine areas in a more sustainable manner. Many of these instruments have a very marginal "political strength" and their implementation is much less than optimal but their mere existence does indicate a change in form of concern and advancement. Chile has a National Fisheries Policy (2007) and a general fishing law, and various regulations on fishing and aquaculture; Peru does not have a comprehensive policy on fisheries and coastal-marine development but a general fishery law and various sectorial regulations; Mexico has a National Fishery Charter (2017) and a general fishing and sustainable aquaculture law. Chile, Mexico and Peru have seen a significant increase in citizen and business initiatives, as well as powerful media campaigns and social network action, which are gradually modifying certain conducts with regards to coastal spaces and seas in general. Clean-up beach campaigns; awareness for the artisanal fishing sector; protection of surf breaks; business initiatives to promote sustainable fishing and catches; systems for fishery quotas; satellite tracking for vessels; marine product certifications; new coastal and marine protected areas, among others, seek to mitigate and revert a long-standing situation related to marine and coastal zones. Although the challenge is huge, the perspectives are very comforting and expectations for change high.

It is against this background that the Peruvian Society for Environmental Law (SPDA) and the David and Lucile Packard Foundation present this publication with a critical policy, institutional and legal overview of significant coastal and marine issues for Latin America and for Chile, Mexico and Peru in particular.

The goal of this publication is twofold: on one hand, it seeks to respond to a need and gap in the reflection and analysis of institutional, policy and legal issues pertaining to seas, oceans, coastal areas and fisheries in Chile, Mexico and Peru; second, it seeks to contribute to policy discussions and national agendas in a context of growing collaboration, opportunities and international funding to promote the construction and implementation of policy and institutional architectures which improve the way in which countries treat their coasts, oceans and resources.

The book consists of 12 chapters, which describe and reflect on a diversity of issues of relevance to Chile, Mexico and Peru, but equally for other countries in Latin America and beyond. It is a compilation of essays with a common structure that offer conclusions and recommendations regarding how to advance and improve conservation and sustainable development of the seas, oceans, fisheries and coastal and marine areas. The publication closes with an epilogue that extracts some of the common lessons and potentially scalable recommendations.

In Chapter 1, "Marine, Coastal and Fisheries Issues in Chile, Mexico and Peru: An Initial Institutional and Policy Review", Bruno Monteferri and Manuel Ruiz of SPDA, present a general overview of the status of marine and coastal areas and existing policy and institutional structures in Chile, Mexico and Peru. Chapter 2, "Industrial Fisheries in Latin America: Challenges and Lessons from Chile, Mexico and Peru" by Santiago de la Puente and Rocío López de la Lama, from the University of British Columbia, reflects on how fishery industries have developed in these three countries, the consequences and some suggestions to improve national public policies. In Chapter 3, "Marine Bioprospecting", Liliana Pardo López of the Institute of Biotechnology of the National Autonomous University of Mexico, describes some advances that have taken place in the field of marine bioprospecting as a promising activity in the area of technological development and research on marine biodiversity with potential in the "life sciences". Chapter 4, "Illegal Fishing and Non-compliance" prepared by Rodrigo Oyanedel of the University of Oxford, reflects on the characteristics of illegal fishing in Chile, Mexico and Peru and the reasons for often limited institutional and legal responses. Andrea Cuba from SPDA, presents Chapter 5, "Extractive Industries in Coastal and Marine Zones", an analysis of the multiple pollution problems facing coastal borders and marine zones due to industrial extractive activities, both of renewable and non-renewable resources. In Chapter 6, "Marine Protected Areas", Pedro Solano and Alfredo Gálvez of SPDA describe different tools from protected areas theory which could be applied to protect coastal and marine spaces, underscoring the process followed to create the National Reserve of the Tropical Sea of Grau in northern Peru, as an example of integration and collaborative efforts by the public sector and civil society. Chapter 7, "Marine and Coastal Planning", by Fernando A. Rosete Vergés of the National Autonomous University of Mexico, reflects on how coastal and marine planning takes place, its enabling conditions and how multiple actors can actively and informedly engage in the process. Luciano Hiriart-Bertrand, José Manuel Troncoso, Carlos Vargas and Alejandro Correa of the NGO Costa Humboldt present Chapter 8, "From Customary Law to the Implementation of Safeguard Measures: the Case of 'Marine and Coastal Areas for Indigenous Peoples' in Chile" which showcases how a specific legal instrument has enabled, through the recognition of rights to coastal indigenous peoples in Chile, the creation of a process which has led to sustainable management of certain fishing coastal areas and fishing zones. In Chapter 9, "Protection of Migratory Marine Species", Ximena Vélez-Suazo of the Smithsonian Institute, describes mechanisms and instruments for the protection of migratory birds and other marine species based on her experiences in Peru. Irene Hofmeijer of Life out of Plastic, in Chapter 10, "Prevention of Marine Pollution from Litter in Peru", analyses ongoing efforts and initiatives undertaken in Peru to combat and mitigate the effects of waste and garbage in the seas, mainly through prevention and recycling activities. Rodrigo Estévez and Stefan Gelcich of the Catholic University of Chile present Chapter 11, "Collective Action Spaces and Transformations in the Governance of Fisheries Resources: Towards Democratic and Deliberative Management", where they reflect on how through social sciences, including economy, law and sociology, major inputs can be identified to develop robust and adequately founded governance systems particularly in the case of common resources. Finally, Bruno Monteferri, Manuel Ruiz and Christel Scheske present Chapter 12, "The Legal Protection of Surf Breaks: An Option for Conservation and Development" where they describe and review new approaches for the preservation and sustainable development of the marine and coastal natural patrimony through the legal protection of surf breaks. A comparative analysis is offered on how these approaches are contributing to nature conservation and appreciation.

These chapters provide a comprehensive overview of a series initiatives, activities and actions that are being undertaken in Chile, Mexico and Peru to address different aspects regarding the rational and sustainable management, protection and development of coastal and marine spaces and their natural resources. For countries like these, with extensive coasts on which many people depend for their survival, the development and application of strategies and respectful measures towards the environment and natural surroundings is critical if these spaces are envisioned as possible options for the generation of wealth, well-being and development for all.

The Peruvian Society for Environmental Law is grateful for the support of the David and Lucile Packard Foundation in the development of this publication.

1 Marine, coastal and fisheries issues in Chile, Mexico and Peru

An initial institutional and policy review

Bruno Monteferri and Manuel Ruiz Muller

Introduction

What do Chile, Mexico and Peru share with regard to their marine and coastal realities and fisheries and their relevant institutional and legal architectures? This chapter explores some common elements that marine and coastal areas and fisheries of these three countries have and the legal and institutional challenges they face for their sustainable development.

Chile, Mexico and Peru have large extensions of coastlines (Chile 6,400 km, Peru 3,100 km and Mexico 11,200 km approximately) and seas (120,000 km^2 Chile, 1,140,000 km^2 Peru and 231,800 km^2 Mexico, approximately). They share a historic tradition of use and occupation of coastal areas, with important urban centers developing rapidly on their shores. These areas contribute significantly to their economies including through small- and large-scale exploitation of fisheries, the presence of a massive off-shore oil industry in Mexico and to a lesser extent Peru and increasingly multifaceted and growing tourism activity in marine and coastal spaces. A large share of urban populations is concentrated on the coastal zones of these countries, especially in Chile and Peru.

These interrelated land and marine areas are equally critical from an ecologic point of view and the services they provide at local, national and global levels. The Humboldt (cold) and Niño (warm) currents that converge in the north of Peru and south of Ecuador are causing climate events with global repercussions. The presence of a type of plankton in marine zones in Chile and Peru makes their seas particularly rich for pelagic fishing. The potential of this resource is still being explored in laboratory conditions (Adiba, *et al.*, 2013). Tourism directed towards the coasts is also growing with large-scale tourist centers and enclaves in each country, mainly Mexico (i.e. Acapulco, Cancun). Even sports such as surfing, wind surfing, underwater spearfishing and body boarding, among others, have grown exponentially over the last two decades and have become an economic support for many local communities in the three countries (Thomas, 2014).

Although archeological evidence shows that fisheries in these countries have been locally relevant long before the Inca, Maya and Aztec periods, it was from the middle of the twentieth century that industrialization gained national and international importance in trade flows (Mann, 2006), and it is only recently that the national political, institutional and legal agendas began to reflect concern for the viability of fisheries in these countries in general, given the intensification in activities.

In the context of marine and coastal zones and the seas, these national agendas in Chile, Mexico and Peru mirror growing international concern for the state of these spaces. In terms of conservation and sustainability, the last two decades, more or less, have seen a sudden reaction from the national and international community concerning the accumulated problems of fisheries overexploitation, coastal pollution, unplanned development of marine and coastal spaces and the effects of climate change on the seas, among others. The Brundtland Report of 1980 marked a milestone in this regard, by calling attention to the state of the global environment around the world (including the oceans) from a multi-sectorial and multidisciplinary approach.[1]

As a result, a number of international conventions (in some cases regional agreements) have been developed and present complex legal architectures that seek to protect ecosystems and marine and coastal species, as well as to prohibit and regulate conducts that are harmful for the marine environment. These conducts include waste disposal, transport of dangerous substances, migratory species, bycatch, etc.[2]

In this context, Chile, Mexico and Peru have privileged coasts and seas which face similar challenges. A look into their realities may offer alternatives and responses potentially scalable to similar realities in each country and beyond. This chapter attempts to identify these challenges and present a few notable examples of legal, policy and institutional experiences and constructs from each country, targeted at sustainably conserving and developing adjacent coasts and seas, as well as fisheries.

Some socioeconomic and environmental achievements

When analyzed quantitatively, the importance of coastal and marine areas and fisheries in Chile, Mexico and Peru is self-evident. Chile is a coastal country due to its geography. The capital Santiago de Chile, less than 120 km away from the Pacific Ocean, concentrates 40 percent of the country's population that amounts to 17.3 million.[3] In the case of Mexico, it is estimated that more than 50 million people live in coastal states. This represents approximately 40 percent of the total population of Mexico.[4] At least 60 percent of the population in Peru lives in large cities on the strip of land along the Pacific. Lima alone, the capital, has 10 million inhabitants, 30 percent of the country's population.[5] The tendency is growing and the phenomenon of populations concentrating in coastline cities is global (Creel, 2003).

All this inevitably implies a direct effect and continuous pressure on the environments and marine and coastal areas of these countries. The impacts from this concentration of population are expressed in numerous ways, from the disposal of wastewater and debris from urban centers into the sea, to the effects of coastal edge infrastructure developments (i.e. piers, housing, highways) on marine biodiversity.

Fisheries for their part are a major contribution to the socioeconomic well-being of Chilean, Mexican and Peruvian societies and to food security for hundreds of thousands of people in local areas. These fisheries, both small/artisanal and industrial, have developed differently in each country. In general terms and concerning catch volumes Chile occupies sixth place in the world, Peru fourth and Mexico sixteenth.[6] Around these numbers there are certainly significant nuances that vary year to year. For example, historically Peru has the largest fishery of a single species on the planet, Peruvian ancho-veta (*Engraulis ringens*), mainly used to produce fishmeal and fish oil and is the first worldwide producer. This fishery alone represents nearly 10 percent of catches of *all* the worlds' fisheries (Heck, 2015).

On the other hand, the population that depends directly and indirectly on fishing and associated activities (i.e. repair services for vessels, restaurants, boat owners, etc.) is equally important. In broad terms, fisheries activities generate 350,000 direct jobs and 2,000 indirect ones in Mexico[7]; 200,000 direct and 800,000 indirect in Peru[8]; while in Chile, 90,000 workers are employed directly and 95,000 indirectly. Despite this, fisheries contributions to the Gross Domestic Product (GDP) continue to be relatively low. In Peru, this contribution fluctuates between 1.5 percent and 2 percent (SNP, 2017); in Mexico the contribution is 0.3 percent (GBC, 2013) while in Chile it reaches 0.4 percent (AQUA, 2019). In general, these are relative low figures when contrasted against contribution to employment and food security.

In addition to this quantitative overview, the value of marine–coastal zones has yet to be calculated in terms of ecosystem services from natural infrastructure and the non–consumptive or material benefits they provide through contemplation, and spiritual, recreational and aesthetic values (Millennium Ecosystem Assessment, 2003). In this respect, Chile, Mexico and Peru have a cultural wealth yet to be enjoyed and benefitted from sustainably.

Legal and policy frameworks

Just as Chile has a National Policy for the Use of the Coastal Border,[9] Peru approved its Guidelines on Integrated Coastal Zone Management,[10] which are in essence, public policies that seek to give these spaces a sustainable treatment. Mexico does not have a general and comprehensive policy or legal framework for its marine and coastal zones but does have focalized policies for specific fisheries and their management. For Chile, Mexico and Peru, infrastructure development, exploitation of non–renewable resources, waste disposal and marine protected areas, among others, as well as fisheries in particular, are governed by specific and sectorial legislation which entails a complex web of interrelations, overlaps and, sometimes, conflicts of competences at different government levels (i.e. central or federal government, municipalities, specialized units, etc.). In other words, the creation and management of marine and coastal protected areas is governed by legislation on protected areas, infrastructure development responds to relevant legislation (i.e. development of roads and highways, or ports and piers, or urban development in general), waste disposal is regulated by an environmental or industrial framework, and so forth. Inter-sectorial coordination among the different levels of government to implement the National Policy in Chile and the Guidelines in Peru, continues to be a challenge and pending matter.

Although there is not a comprehensive and systematic policy or law to address or integrate the different dimensions of fisheries with marine and coastal related issues, it is important to emphasize that in the biodiversity strategies and action plans of Chile, Mexico and Peru, references to biodiversity conservation and marine and coastal ecosystems have been made either at the species or ecosystem levels. These policy instruments are an important, albeit often overlooked, references to inform both regulatory actions as well as specific interventions in these areas. For example, Chile has developed a National Biodiversity Strategy 2017–2030 (2016) that includes a detailed assessment on the situation of marine and coastal ecosystems and oceanic islands; it proposes measures to integrate marine biodiversity in sectorial policies, plans and programs and develops a thematic approach to marine biodiversity and islands.[11] In the case of Mexico, the National Biodiversity Strategy and Action Plan 2016–2030 (2016)[12] also has explicit references and actions aimed at restoring vulnerable marine and coastal ecosystems; ensuring the continuity of ecosystem biogeochemical processes in infrastructure planning in coastal and island areas, and generating incentives for community participation in the restoration of marine–coastal ecosystems in terms of their environmental services. Finally, in the case of Peru, the National Strategy for Biological Diversity 2021 and Action Plan 2014–2018,[13] does not include major references to marine and coastal areas, except with regards to the need of establishing a type of sustainable management and operating modality for at least 10 percent of the biodiversity in marine areas and some action in terms of valuation, education and awareness on marine and coastal biodiversity. The focus of these strategic instruments is, primarily, on better understood continental and terrestrial biodiversity. None of the three countries has a specific strategy dedicated to marine and coastal biodiversity. The exception may be Chile to some extent due to its location and particular geography.

In terms of marine and coastal protected areas, advances in Chile, Mexico and Peru are quite dissimilar. Nearly 42 percent of the marine territory in Chile is under a form of special protection or management. It has 33 recognized protected areas in marine zones.[14] Chile has also developed the Marine and Coastal Areas for Indigenous Peoples, a new category, different to a protected area but with similar features, that proposes exclusiveness in the management and administration of marine–coastal ecosystems where ancestral communities of fishers have historically developed.[15] Mexico for its part has 37 marine-coastal protected areas that cover 22 percent of the marine territorial surface, protected under some specific category.[16] Both Mexico and Chile have considerably exceeded their commitments to the Aichi Biodiversity Targets to protect

10 percent of their sea territory. Beyond the forever-existing limitations to guarantee management and monitoring mechanisms for the protection given to these spaces, there is evidence of political commitment, mainly in recent years. Peru is still far behind in terms of marine protected areas, with only 0.5 percent of its marine territorial extension protected through four natural areas dedicated to the conservation of marine and coastal spaces and biodiversity.[17]

In the case of fisheries, unlike Mexico and Peru, Chile has a National Fisheries Policy (2007)[18] and a National Aquaculture Policy (2003)[19] which despite the time lapsed, show certain clarity in relation to the order in which the activities must be approached in national development agendas. Additionally, Chile has a General Fisheries and Aquaculture Law (updated and modified in 2017) and complementary regulations, which include modern and innovative approaches related to inspection, fisheries quotas and protected zones, among others.[20] Mexico has a General Law for Sustainable Fisheries and Aquaculture from 2007 (modified in 2018) that also includes innovative aspects such as a National Fisheries Chart, a national fund for sustainable fishery and aquaculture, differentiated instruments for small and large operators, a detailed sanctions regime and a National Fishery Council as a multi-sectorial counseling space for fisheries and aquaculture. Finally, Peru has a General Law on Fisheries from 1992, that has been modified over time on specific aspects, such as creating subsequent regulations with regard to fisheries quotas, differentiated regimes for artisanal and industrial fisheries. One of the advances made over time under the Law was the adoption of a framework for fisheries quotas.[21] Due to the time passed, the General Law on Fisheries requires a reformulation in order to adapt to the modern realities of fishery resources management, conservation and sustainability at all levels.

In institutional terms, there are some similarities and differences between Chile, Mexico and Peru. Chile has a Sub-Secretariat of Fisheries and Aquaculture, which depends on the Ministry of Economy, Development and Tourism, as the maximum authority responsible for determining public policies in the context of fisheries and aquaculture. In Mexico, this responsibility falls at the federal level, on the Secretariat of Agriculture, Livestock, Rural Development, Fishing and Food. In Peru, the Ministry of Production, through the Vice-Ministry of Fisheries, determines fisheries public policies. These are the policy-setting bodies that set the direction for countries in terms of fisheries and aquaculture policies. However, within the fisheries sector in general, there are many other public institutions with competences regarding funding, promotion, inspection and research, etc.

With regards to fisheries innovation and research to inform policy making, the Institute of the Sea of Peru is responsible for generating and providing the Vice-Ministry of Fisheries and other actors, data and technical and scientific information on the status of the country's fisheries. The National Institute of Fisheries and Aquaculture of Mexico is a decentralized and public body which is in charge of directing, coordinating and guiding scientific and technological research in terms of fisheries and aquaculture, as well as the development, innovation and technological transfer required by the fisheries and aquaculture sector. In the case of Chile, the Institution for Fisheries Development undertakes technical and scientific functions, orienting their actions towards generating, developing and transferring valuable knowledge to provide the authority with necessary information to permanently evaluate the state of resources that sustain major fisheries.[22]

Common challenges and obstacles for the three countries

There are three major challenges that Chile, Mexico and Peru face in the years to come. First, the *implementation* of multiple instruments and tools already in existence is essential, to promote sustainable development in marine and coastal zones and ensure the viability of their rich fisheries. As in other Latin American countries and environments, many times it is not a question of lack of norms or policies but rather of their concrete application. This demands leadership and long-term commitments in order to convert marine–coastal spaces and national seas into driving forces for social and sustainable development.

Second, it is important to generate an *awareness* process among society in order to understand that the sea and marine and coastal zones are not only areas for the exploitation of natural resources or amusement. These areas provide multiple environmental services and contribute significantly to food security for populations, although their effect could be improved through better strategic planning. If you live "with your back to the sea" it is now necessary to integrate it into national development.

Finally, it is also necessary to substantially improve *control, inspection* and *sanctioning* actions for the multiple irregular and openly illegal activities undertaken on the coast and sea. Building infrastructure without adequate environmental assessments, illegal fishery both artisanal and industrial, pollution of marine and coastal zones due to industrial waste, and others, are only some of the problems that regularly emerge in Chile, Mexico and Peru. Impunity and corruption, especially in Mexico and Peru, undermine good governance and administration of vulnerable spaces that require special attention for their sound management and use.

Actions on these three fronts could substantially contribute to improving the quality of marine and coastal spaces and the seas. Such measures would also contribute to the international front in terms of compliance with the Sustainable Development Goals and Aichi Biodiversity Targets.

Final reflections

How to integrate different aspects and dimensions related to fisheries and marine and coastal spaces in a conservation and sustainable development approach, is one of the major challenges for countries. Due to its specific characteristics in geographical terms, Chile has taken great strides to integrate their coasts and marine zones to their development approach. In spite of their equally extensive coasts, Mexico and Peru have only recently started to consider the spaces of sea and marine and coastal zones not only for the extraction of natural resources but also as important assets for national development.

In recent years there have been several initiatives by civil society to promote actions intended to improve environmental conditions for the coasts and seas. Campaigns against the use and disposal of plastics, initiatives to clean the beaches, campaigns directed to the sustainable consumption of marine species, campaigns for the protection of surf breaks or adjacent spaces, among others, have begun to raise awareness and sensitize the users of coastal zones and seas to their crucial importance and also their vulnerability.

A situation in which Chile, Mexico and Peru look at the sea without the idea of *only* extracting and are more sensitive to the important role marine and coastal spaces play in national development is perfectly foreseeable in the medium term.

Notes

1 The history of environmental law can be divided into three great moments: the 1960s and 1970s, with the growing scientific concern for environmental problems and the Stockholm Conference on the Human Environment (1972) and the first multilateral environmental conventions; the process of Our Common Future (Brudtland Report) in the 1980s, that highlighted global nature to environmental problems, including the loss of biodiversity and fisheries; and the process of the United Nations Conference on Environment and Development (1992), that strengthened international action resulting in another significant set of international conventions.

2 Some of the most recognized agreements include the Convention on Biological Diversity (CBD, 1992), the FAO Code of Conduct for Responsible Fisheries (1995), the United Nations Convention on the Law of the Sea (1982), the Convention on the Conservation of Antarctic Marine Living Resources (CCRVMA, 1980) and the Convention on the Conservation and Management of High Seas Fisheries Resources in the South Pacific Ocean and the Convention on Wetlands of International Importance Especially As Waterfowl Habitat (RAMSAR, 1972).

3 See, www.emol.com/noticias/Nacional/2017/08/31/873382/Resultados-preliminares-del-Censo-2017-Poblacion-en-Chile-llega-a-17373831.html.

4 See, https://noticieros.televisa.com/historia/mares-y-costas-de-mexico-asi-de-importantes-son-para-la-economia/.

5 See, https://es.wikipedia.org/wiki/Demografía_del_Perú.

6 *Top 10 Largest Fish Producing Countries in the World* (2018), available at, www.worldblaze.in/largest-fish-producing-countries/.

7 See, www.excelsior.com.mx/nacional/mexico-se-ubica-como-lider-en-produccion-pesquera-en-al/1240441.

8 This data varies considerably according to sources. There is a certain coincidence in the sense that direct jobs reached between 200,000 and 220,000. The measuring of indirect jobs is much more complex. See for example: *El Peruano*, February 20,

2019, available at https://elperuano.pe/noticia-la-exportacion-pesquera-no-tradicional-sumara-1539-mllns-56420.aspx.

9 National Policy for the Use of the Coastal Border (1994), in control of the Ministry of Defense, available at www.ssffaa.cl/asuntos-maritimos/cnubc/politica-nacional-uso-del-borde-costero/. Even in the case of Chile, this National Policy must be read systematically with other specific legal instruments that address particular marine and coastal zone issues.

10 Approved by Ministerial Resolution 189-2015-MINAM, of August 5, 2015.

11 National Biodiversity Strategy 2017–2020. Available at http://portal.mma.gob.cl/wp-content/uploads/2018/03/Estrategia_Nac_Biodiv_2017_30.pdf.

12 Available at www.biodiversidad.gob.mx/pais/enbiomex/pdf/ENBIOMEX_baja.pdf.

13 Approved through Supreme Decree 009-2014-MINAM, on November 5, 2014, available at file:///C:/Users/MANOLO/Downloads/diversidadbiologica.pdf.

14 See, www.biologiachile.cl/2018/06/12/chile-pionero-en-proteccion-de-su-biodiversidad-marina/.

15 Law 20.249 of February 16, 2008, that creates the Coastal Marine Areas for Indigenous Peoples.

16 See, www.gob.mx/conanp/prensa/mexico-es-lider-mundial-en-la-proteccion-de-areas-marinas.

17 As part of the Director Plan for Protected Areas work has intensified and attention focused on the potential of natural protected areas in the marine and coastal environment.

18 National Fisheries Policy (Chile), available at www.diputados.gob.mx/LeyesBiblio/pdf/LGPAS_240418.pdf.

19 National Aquaculture Policy, available at www.subpesca.cl/portal/616/articles-60019_recurso_5.pdf.

20 Law 21.033, General Law for Fisheries and Aquaculture.

21 The Law on Maximum Catch Levels per Vessel ("Ley de Cuotas de Pesca") was enacted through Legislative Decree 1084, of June 28, 2008 and its regulation through Supreme Decree 021–2008-PRODUCE, of December 11, 2008.

22 See, www.ifop.cl/nuestro-que-hacer/la-investigacion-pesquera/.

References

Adiba, H., Ruchaud, S., Rios, L., Hummeau, A., Probert, I., De Vargas, C., Bach, S., and Bowler, C. (2013), Bioprospecting Marine Plankton. *Marine Drugs.* 11(11): 4594–4611. Available at www.ncbi.nlm.nih.gov/pmc/articles/PMC3853748/.

AQUA (2019), *En Chile: La Pesca Garantiza Mucho más que la Subsistencia.* Available at www.aqua.cl/2016/02/01/en-chile-la-pesca-garantiza-mucho-mas-que-la-subsistencia/.

Creel, L. (2003), *Ripple Effects: Population and Coastal Regions.* Population Reference Bureau. Available at www.prb.org/wp-content/uploads/2003/09/RippleEffects_Eng.pdf.

GBC Group (2013), *La Industria Pesquera en México,* available at www.gbcbiotech.com/genomicaypesca/pesca_en_mexico.html.

Heck, C. (2015), *Hacia un Manejo Ecosistémica de la Pesquería de Anchoveta.* SPDA. Lima, Perú. Available at https://spda.org.pe/wpfb-file/anchoveta-pdf/.

Mann, C. (2006), *1491: New Revelations about the Americas Before Columbus.* Penguin, Random House. London.

Millenium Ecosystem Assessment. (2003), Ecosystems and Human Wellbeing. Synthesis

Report. Available at www.millenniumassessment.org/documents/document.354. aspx.pdf.

SNP. (2014), *Aportes al Debate de Pesquerías* (No. 1). Available at, https://snp.org.pe/media/pdf/aportes-al-debate-en-pesqueria/2014-02-24-SNP-Aportes-al-Debate-Relevancia-Economica-del-Sector-Pesquero.pdf.

Thomas, G. (2014), Surfonomics Calculates the Worth of Waves, *Washington Post*, August. Available at www.washingtonpost.com/surfonomics-quantifies-the-worth-of-waves/2012/08/23/86e335ca-ea2c-11e1-a80b-9f898562d010_story.html?utm_term=.d13b30f01cf6.

2 Industrial fisheries in Latin America

Challenges and lessons from Chile, Mexico and Peru

Santiago de la Puente and
Rocío López de la Lama

Introduction

Multiple types of fisheries have developed in the planet's marine ecosystems. Industrial fishing, the subject of this overarching chapter, operates on a large scale and for commercial purposes, it requires important capital investments, uses technology to facilitate detection and extraction of resources, takes place in coastal as well as open sea areas and involves crews and vessels that schedule fishing trips that may last days or even months (Panayotou, 1983; Teh and Pauly, 2018).

Industrial fishing extended globally during the decades following the post–Second World War era, expanding geographically (from the northern to southern hemisphere), bathymetrically (from the surface to the deep sea) and taxonomically (gradually introducing a larger number of species) (Pauly, 2009). This activity has continued to grow for more than 70 years and is responsible for worldwide catches of fish and marine invertebrates. However, its total contribution to fish production has steadily declined since 1996 (Pauly and Zeller, 2016). To understand this trend, a historical perspective is merited.

During the early days of the industrialization of fishing, this economic activity was like any other, where greater capital investments (e.g. more or larger vessels and improved systems to detect schools of fish) resulted in an increase of production (Pauly, 2009). However, the increase of capacity and fishing efficiency became excessive. This led to an imbalance between increased captures and the regenerative capacity of species, which resulted in the collapse of numerous fisheries, negatively impacting marine ecosystems and social systems that depended on them (Myers and Worm, 2003; Daly, 2005; Worm et al., 2006; Halpern et al., 2008).

At the same time, the collapse of fisheries during the second half of the 20th century was also critical to the redesign and improvement of management systems (Ferguson-Cradler, 2018), whose implementation brings us closer to the sustainable development of marine resources (Mangin et al., 2018).

At present, the targets of industrial fishing can be divided into two large groups: overexploited (lower productivity) and sustainably used species. In this context, industrial fishing faces three challenges to ensure its development in perpetuity by ensuring timely recovery of overexploited stocks. These refer to, first, guaranteeing sustainable management of targeted resources, second, minimizing negative ecosystem impacts from its activities, and third, strengthening fisheries governance.

These challenges are not exclusive to industrial fishing but given the magnitude of catches and its socioeconomic importance, an analysis and reflection is required. To do this, the chapter first describes industrial fishing activities in Chile, Mexico and Peru; it then details how these countries are addressing the challenges, identifying lessons learned also applicable to other countries in the region; finally, critical issues related to management and sustainability are presented and discussed.

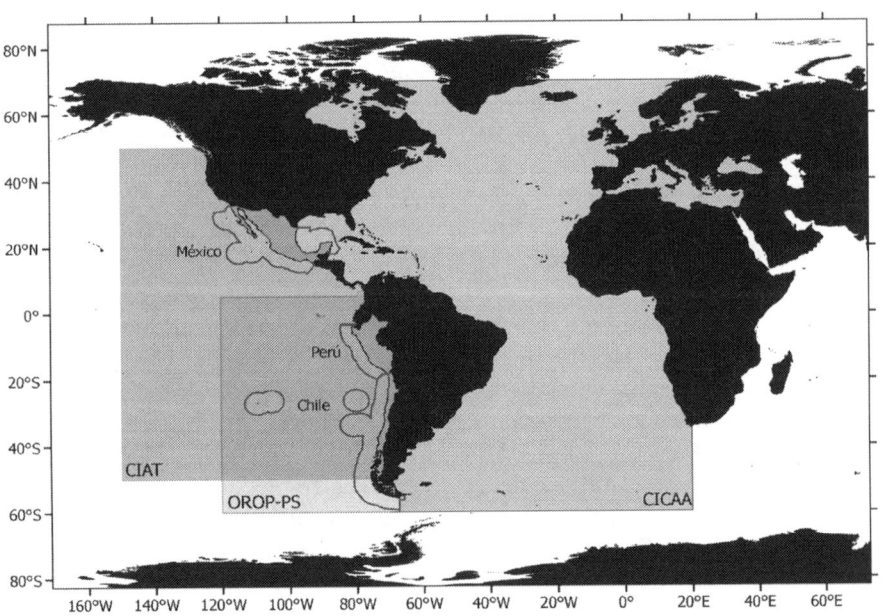

Figure 2.1 Fishing areas of Chile, Mexico and Peru. The adjacent areas on the coastal borders of these countries represent their exclusive economic zones. Delimited areas outside these zones are regulated by regional fisheries management bodies. Chile and Peru undertake industrial fishing outside their economic zones, in areas governed by the South Pacific Regional Fisheries Management, as well as by the Inter-American Tropical Tuna Commission. Mexico fishes outside its exclusive economic zones in areas governed by the Inter-American Tropical Tuna Commission and by the International Commission for the Conservation of Atlantic Tuna.

Table 2.1 Definition of industrial fishing vessels according to the country

Country	Definition	Source
Chile	Boat lengths exceeding 18 meters, with technological systems for fishing, such as trawling, longline and nets, which allows massive catches of a wide variety of fish.	Article 2: New Fishing Law No. 20,657 of 2013.
Mexico	Boat lengths exceeding 10.5 meters, equipped with a stationary engine, a continuous bulkhead deck, areas to maneuver the fish, navigation and communication equipment, eco-detection and satellite tracking, equipped with cabins, a kitchen and bathroom, where the conservation of products can be with ice or refrigeration systems with an average autonomy of 20 days.	Article 4: General Fishing and Sustainable Aquaculture Law (2007).
Peru	Boats with storage bulks exceeding 32.6 cubic meters.	Article 20: Regulation of the Fishing Law Supreme Decree, No. 012-2001-PE.

Industrial fishing in the Chilean, Mexican and Peruvian context

Generalities and numbers

Chile, Mexico and Peru are the industrial fishing leaders of Latin America and capturing fish and marine invertebrates within and outside their exclusive economic zones (Figure 2.1). These countries classify industrial fishing based on the technical and physical features of fishing vessels, which vary from country to country (Table 2.1). Purse seines and trawl nets are the most utilized methods of catch by their industrial fishing fleets (Cashion *et al.*, 2018a).

Industrial fishing in these three countries is the main form of extraction of marine resources. Both in Chile and Peru, this activity was responsible for 87 percent of total captures in each country between 1950–2014 (Mendo and Wosnitza-Mendo, 2014; van der Meer *et al.*, 2015). In Mexico this number represented only 57 percent of total captures during that period (Cisneros-Montemayor *et al.*, 2015).

Industrial captures in each of these countries exceeds 1 million tons per year. The Peruvian industrial fishing sector is the largest of the three in terms of production. During 2010–2014, Peru reported average yearly captures of 6.13 million tons (Mendo and Wosnitza-Mendo, 2014). During the same period, Chile and Mexico reported average yearly captures of 2.47 million tons and 1.24 million tons respectively (Cisneros-Montemayor *et al.*, 2015; van der Meer *et al.*, 2015).

Of the three countries, the industrial fisheries sector of Mexico has the most diverse portfolio of target species. They capture small pelagic fish such as sardines (*Sardinops sagax*), northern anchovy (*Engraulis mordax*) and Pacific thread herring (*Opisthonema libertate*); tunas such as yellowfin tuna (*Thunnus albacares*) and skipjack tuna (*Katsuwonus pelamis*); crustaceans such as blue shrimp (*Litopenaeus stylirostris*) and northern brown shrimp (*Farfantepenaeus aztecus*) (Arreguín-Sánchez and Arcos-Huitrón, 2011; Cisneros-Montemayor *et al.*, 2015). Historically, their main industrial catches have been sardines, but they do not exceed 30 percent of total captures reported between 1950–2014 (Cisneros-Montemayor *et al.*, 2015).

On the other end of the spectrum, industrial catches in Peru concentrate almost exclusively on Peruvian anchovy (*Engraulis ringens*) – a species that represents 89 percent of the total industrial catches (Mendo and Wosnitza-Mendo, 2014; Cashion *et al.*, 2018a). Chile is mid-point, mainly focusing its industrial fleet on fishing four species, the most important being Chilean jack mackerel (*Trachurus murphyi*) and anchovy (van der Meer *et al.*, 2015). Unlike Mexico, Chile and Peru dedicate most of their catches to the production of fishmeal and fish oil for export purposes (Cashion *et al.*, 2017).

The industrial fisheries sectors of these three countries are important employment sources and income generators. For example, the sector in Chile employs around 3,500 industrial fishers and generates close to 26,502 jobs, including the industrial facilities and processing plants (SUBPESCA, 2018). In 2012 this sector contributed an estimated 4.5 percent to the Chilean GDP (Miller, 2014). Similarly, this sector in Peru employed 49,000 people during 2009 and was responsible for 41 percent of the fisheries GDP, producing US $3.43 billion (Christensen *et al.*, 2014a). In Mexico, the contribution of industrial fisheries to employment and the sector's and national GDP is less visible given the way in which national statistics are reported.

Industrial fisheries management

Fisheries management is a comprehensive process that involves the collection of data and analysis, as well as planning, consultation and decision-making (Cochrane, 2002). To improve productivity and achieve management object-ives, the process also involves fishing regulations, including the assignment of rights, as well as development of tax and criminal norms.

To prevent and mitigate overfishing, management systems rely on various tools. These include establishment of *output limits* which prevent excessive catches, and *input limits*, which restrict the total amount of fishing activities in a determined area for a fixed period of time (Walters and Martell, 2004).

The most common mechanism to regulate output limits globally are general catch quotas (Caddy, 1999). This requires state or regulatory bodies to establish a limit to extraction of a particular resource in a determined area over a period of time. On the downside, general quotas lead to increases in the power of

Table 2.2 Number of fisheries with systems of general and individual quotas according to each country

Country	General quotas	Individual quotas	Industrial individual quotas
Chile	17	14	13
Mexico	17	9	0
Peru	11	2	2

Data source: EDF (2018).

fishers and displacement of competition. They make investments to expand their fleet with larger and better-equipped vessels. As a result, fishing vessels grow excessively (fishing overcapacity) and progressively capture their quotas in a shorter period of time (race for fishing). This situation results in additional costs that end up encouraging fishers to catch over established limits, which eventually leads to overfishing (Anderson and Seijo, 2010).

To mitigate this problem, general quotas are complemented with regulations that also place restrictions on fishing (Walters and Martell, 2004). In Chile, Mexico and Peru, most industrial fisheries have general fishing quotas plus other output limits such as minimum catch sizes and tolerance percentages for juvenile fish catch. Additionally, they also have in place regulations for input control such as licensing or permit requirements to access fisheries, temporary restrictions on fishing (i.e. closure periods), areas banned for industrial fishing and regulations regarding the characteristics of catching methods for this industry.

Even with these systems in place, many marine resources in these countries are overexploited or have collapsed (INAPESCA, 2018; SUBPESCA, 2018), due to overcapacity and the race for fishing (Castilla, 2010; Tveteras *et al.*, 2011). To address this problem and following global trends, Chile, Mexico and Peru have implemented systems for *individual* fishing quotas (Table 2.2). These are schemes where a vessel, a vessel owner or group of owners, obtain exclusive access, during a period of time, to a fraction of the general fishing quota (Caddy, 1999). In theory, by granting exclusive access rights to users incentivizes them to protect the resources they exploit, given overfishing directly affects their interests. This situation would not occur under alternative schemes, where they have no guarantees to future benefits, triggered by management actions that restrict their present catches (Anderson and Seijo, 2010).

Although individual quotas are not the panacea for fishing-related problems (Walters and Martell, 2004; Anderson and Seijo, 2010), Chile began to implement them in 1991 when the Fishery Law entered into force.[1] According to this law, access rights to fisheries are transferable, divisible and independent from the vessels used for fishing. The initial assignment of such rights was based on historic catches of each vessel and storage capacity. At present, access to these or new fisheries follows an auction system. The new Fishery Law modifies previous legislation in favor of sustainability principles

(BCN, 2014).[2] Furthermore, it also recognizes the importance of resource conservation, recognizes the precautionary principle as a guide for sustainable fisheries policy and provides for the creation of Fisheries Scientific Committees. There are eight committees that operate as advisory bodies to define the status of fisheries, identify biological reference points for resource management, determine the range for general quotas and propose biologically acceptable quotas.

On the other hand, the first Federal Fishing Law of Mexico was enacted in 1986; it was modified in 1992 and 2001 and later replaced by the General Law for Sustainable Fishing and Aquaculture (2007, amended in 2018). The law is yet to be regulated and is applied through a 1999 regulation (modified in 2004). The importance of the law is that for the first time, fishing and aquaculture activities are recognized as a national security issue for Mexico. Likewise, responsibilities in sectorial development plans, and inspection and monitoring functions, are granted at federal, state and municipal levels (Alfaro and Quintero, 2014). However, exclusive access rights or individual fishing quotas for industrial vessels are not assigned through this or any other Mexican regulation.

The General Law of Fishing regulated activities in Peru since 1971. It has been modified on various occasions and replaced entirely twice (Arias-Schreiber, 2012). The present General Law of Fishing[3] was enacted in 1992. Its regulation has also suffered multiple modifications.[4] It was enacted in 2001 and last modified in November 2018.[5] The law and regulation seek to regulate fisheries to promote sustainable development and the responsible use of fisheries resources, whilst recognizing that fishing is a key activity for food security, employment and income (Heck, 2015). Only two fisheries are subject to individual fishing quotas: Peruvian hake (*Merluccius gayi peruanus*)[6] since 2003 and anchovy for indirect human consumption since 2009.[7]

It is important to note that the institutional framework governing industrial fishing management is similar in the three countries. Ordering falls mostly on second- and even fourth-level institutions of the executive branch. For example, in Chile, the administrative and regulatory responsibility for industrial fishing fall under the National Sub-Secretariat of Fishing and Aquaculture. The National Service for Fishing and Aquaculture is responsible for gathering information on landings and undertaking inspections. Both institutions are ascribed to the Ministry of Economy, Development and Tourism. In Mexico, the National Commission on Aquaculture and Fishing is the public entity responsible for designing and implementing fishing policies, as well as ensuring compliance with the law. The National Commission depends on the Secretariat of Agriculture, Livestock, Rural Development, Fishing and Food. In Peru, the Ministry of Production is responsible for regulating and managing the fishery sector, including by imposing sanctions. Under the Regulation on Organization and Functions of the Ministry, all fishing and aquaculture activities are under the responsibility of the Vice Ministry of Fishery.[8]

Research on and monitoring and evaluation of uses of marine resources are critical areas for fisheries management and planning. In Mexico and Peru these responsibilities fall on state agencies. The National Institute of Fishing and Aquaculture provides the National Council on Fishing and Aquaculture with scientific data and information required for the decision-making processes in the Mexican fisheries sector. Likewise, the National Institute of the Sea is responsible for providing the Ministry of Production with scientific information to develop and implement fisheries management measures. In Chile, however, this scientific function falls on a private non-profit organization with a public role: the Institute for the Promotion of Fishery. The Institute undertakes research required to support the management of fisheries and actively engages in Chilean decision-making processes (BCN, 2014).

Main challenges facing Chilean, Mexican and Peruvian industrial fisheries

Towards the sustainable management of target resources

To ensure sustainable management of target resources in industrial fishing, requires regulatory authorities and states level or regional fishery management organizations to oversee that exploitation rates do not exceed their regeneration capacity, by adopting a precautionary approach to risks in the face of uncertainty (Walters and Martell, 2004). This is critical, given the effectiveness of management measures, as well as individual fishing quotas, depend on general quotas being defined correctly (Anderson and Seijo, 2010).

Fisheries that develop in marine ecosystems with highly variable oceanographic conditions demand adaptive management tools. Therefore, fishing quotas in Chile, Mexico and Peru should be determined according to extraction rates, which vary depending on the state of the ecosystems and biomass of the exploited resource, using biological reference points (Walters and Martell, 2004). Fishing quotas must not be defined based on static values such as maximum sustainable yields, whose consistency is greater in stable ecosystems, but which have limitations known and discussed since the 1970s (Larkin, 1977). Extraction rates (adaptive) must be accompanied by measures that limit access and fishery efforts, as well as prevent the capture of juvenile individuals (minimum sizes) and protect spawning and fish breeding areas (fisheries exclusion zones) (Anderson and Seijo, 2010).

A successful example is the management of the anchovy stock in the north-center of Peru. This management is highly adaptive, uses biological reference points for the definition of quotas, applies the precautionary principle in light of potential changes in the state of the marine environment and relies on measures that limit access to fisheries (Oliveros-Ramos and Díaz Acuña, 2015).[9] Furthermore, the state of this anchovy stock is healthy, and the industrial fishery activity can be considered sustainable (Cashion *et al.*, 2018b). Additionally, this fishery relies on an efficient monitoring system, as well as continued assessments regarding the state of the resource.

Fisheries policies must have explicit and quantifiable management object-ives, for their performance to be monitored, modeled and evaluated, recog-nizing mutual concessions needed between objectives defined and their impact on actors involved in the fishery (Walters and Martell, 2004). In this regard, Chile and Mexico have made more progress than Peru. The two countries administer their fisheries based on management plans that includes goals and objectives. Peru on the other hand, has Fisheries Management Regulations that synthesize management tools but lack measurable goals and objectives that would allow for effectiveness to be measured (Heck, 2015).

Countries must also avoid practices that obscure the reality of the state of industrial fisheries. For example, Mexico and Chile continuously publish evaluations on the state of their fisheries on the official websites of the National Commission for Fishing and Aquaculture and the Sub-Secretariat for Fishing and Aquaculture, respectively. These publications show when resources are in a state of overexploitation, depletion or collapse. In contrast, the state of fisheries in Peru is not made public. Moreover, those publications which recognize that the exploited stocks have deteriorated pass from '*fully exploited*' fisheries to fisheries '*in recovery*', without necessarily requiring changes in their Fishery Management Regulation (Heck, 2015).

Industrial fisheries in Mexico are heavily subsidized, with almost US $73 million a year directed to reduce the cost of the vessels' fuel (Cisneros-Montemayor *et al.*, 2016). This has resulted in overfishing of resources such as sardine and shrimp, where the economic profit from industrial fishing would be negative without subsidies (Cisneros-Montemayor *et al.*, 2013). This mechanism produces present (e.g. public funds assigned to subsidize fisheries) and future economic losses, as it prevents the recovery of overex-ploited stocks (Pauly, 2009; Anderson and Seijo, 2010).

Towards fisheries management with an ecosystem approach

Fisheries cannot be separated from the ecosystems on which they depend and in which they develop. Extracted resources play various ecological roles by interacting with other species. Therefore, their extraction can produce neg-ative ecological impacts by affecting the structure and function of marine ecosystems. This may also have negative socioeconomic impacts by directly or indirectly affecting income of actors that exploit stocks, predators and competitors of the target resource (Pikitch *et al.*, 2014; Christensen *et al.*, 2014b). This is particularly relevant in Latin American given industrial and artisanal fishing regularly compete for the target resources and the extraction of resources by one may indirectly impact the other.

The response to this situation is to adopt fisheries management with an ecosystem approach. This means incorporating considerations regarding the impact fishing activities have on the ecosystem, not only on the target resource, in decision-making processes. This can be done by applying complex ecological models where the impact is directly evaluated (Christensen *et al.*, 2014b). For example, it was determined that in the Barents Sea (Arctic Ocean) the main Atlantic cod prey (*Gadus morhua*) was the caplin (*Mallotus villosus*) (Olsen *et al.*, 2010). Both the cod and caplin are target species for different industrial fisheries, but to prevent the collapse of cod fishing, caplin fisheries close when their biomass falls below 200 thousand tons (Pikitch *et al.*, 2014). At present, not one industrial fishery in Latin America has a similar fishery management mechanism, even though small pelagic fish (e.g. anchovies and sardines) exploited by industrial fisheries in Chile, Mexico and Peru, serve similar ecological roles to those of caplin fish (Pikitch *et al.*, 2014).

It is important to note that using ecological models to assess the ecosystem impact of fisheries is a demanding process that requires abundant information, as well as a critical mass of duly trained experts to undertake the task. It is therefore also critical that research institutions supporting fisheries management at the national and regional levels are properly funded, as well as provided with continuous training for their personnel. Nevertheless, there are digital repositories with multiple complex ecological models.[10] These have been published in specialized scientific magazines and can be downloaded freely and updated, with new information available, or adapted to seek the response to other research questions.

In order to minimize the negative impacts from industrial fishing on marine ecosystems, it is also important to prevent and mitigate the incidental catch of species – accidental catches of species other than the target species (Rochet *et al.*, 2014). Unfortunately, selectivity of the main industrial fishing methods in Latin America is limited. For example, industrial trawling in Mexico captures a ton of shrimp for every 10 tons of bycatch, and the country still lacks tangible measures to prevent and mitigate these impacts (Cisneros-Montemayor *et al.*, 2013).

Countries like Canada and New Zealand have successfully managed reductions in bycatch in their fisheries, by applying individual transferable quotas systems for bycatch (O'Keefe *et al.*, 2014). In these cases, vessels that exceed their individual bycatch quotas before capturing their individual quota of the target resource are prevented from continuing their fishing operations. As a consequence, they either stop fishing operations or proceed to purchase the bycatch quota from other vessels. This system has encouraged vessels to be more selective and not to exceed the total bycatch quota (O'Keefe *et al.*, 2014).

Nevertheless, policies like these require a strong rule of law to stimulate commitments by users and important public investments to monitor compliance with individual bycatch quotas. This may limit effectiveness in the present context of Latin America.

Alternatively, Chilean and Peruvian legal frameworks have measures to reduce bycatch based on the zoning of fishing grounds. Both countries impede industrial fishing within the first 5 miles of their Exclusive Economic Zones. This is due to the poor selectivity of purse seines and industrial trawls, and the high ecological value of these coastal border areas for fish and invertebrate breeding and reproduction (Seitz *et al.*, 2014). Defining this exclusion area has a double function: to reduce the probability of bycatch, mainly of species exploited by artisanal fishing, and prevent fishing nets from being dragged across the seabed, endangering associated biodiversity. Chile furthermore prevents the development of industrial bottom trawling activities on vulnerable marine ecosystems that need special protection, such as seamounts.

Peru has decided to include tolerable percentages for bycatch in its Fisheries Management Regulations. In theory, these could be used to temporarily close zones where bycatch that exceeds the tolerable percentage is being reported, or to close a whole fishery if vessels register bycatch above the product of multiplying the tolerable percentage by the quota of the target resource. However, there is no evidence that these percentages have been used as reference points for fisheries management or the prevention of bycatch. They are used alternatively to sanction vessel landings of 'non-target resources' that exceed tolerable percentages.[11] This situation may be encouraging the discard of species, a practice of simply dumping dead or live unwanted catches into the ocean.

Discarding masks the bycatch and prevents an adequate quantification of the fishery's impact on marine ecosystems (Pauly, 2009; Pauly and Zeller, 2016). To counter this problem, Chile has enacted six management plans that seek reductions in discard by different industrial fisheries (SUBPESCA, 2017). These management plans also encourage the implementation of fishing practices to prevent or reduce bycatch, placing Chile among the countries most committed to the reduction of bycatch in Latin America (Sapag *et al.*, 2016).

Towards sound fisheries governance

Sound governance is one of the most important factors for the effective, fair and responsible management of natural resources and in this particular case, to guarantee sustainable management of targeted resources by industrial fisheries and minimize their potential negative impact on the ecosystems. Governance is understood not as a synonym for government, but as the interaction and collaboration among governing institutions and citizens in decision-making processes (Graham *et al.*, 2003).

Good governance is effective, equitable, proactive, robust, dynamic and adaptive (Bennett and Satterfield, 2018). In the context of fisheries, this requires (De la Puente *et al.*, 2011):

1   A legal framework that explicitly incorporates management objectives, detailing decision-making processes (i.e. how to define a catch quota or assign a portion of the quota to a vessel or shipowner);

2   Solid institutions whose roles have been clearly defined, with spaces for citizen participation and mechanisms for conflict resolution;

3   Supervision and sanction systems that ensure implementation of rules (i.e. prevent systematic and repetitive breaking of the law) and learning by institutions in charge of fisheries management;

4   Continuous research to address the need for information and thus reduce uncertainty; and

5   Timely monitoring and evaluation systems to adapt the legal and institutional frameworks related to management.

The three countries referred to in this essay are taking important steps to strengthen their fisheries governance. Chile has probably advanced the most through its new Fishing Law which defines management objectives and planning mechanisms and creates Fisheries Scientific Committees and Management Committees.

Fisheries Scientific Committees are participative spaces created to determine the state of fisheries, biological reference points and ranges in which the general catch quotas can be determined (BCN, 2014). The idea behind this is to separate the scientific processes associated to fisheries management from political pressures, often steered by employment or economic growth factors.

On the other hand, Management Committees are consultative and advisory bodies for the fishery authority and formed by sectoral representatives of each fishery (industrial or artisanal) and of the National Sub-secretariat of Fishing and Aquaculture and the National Service for Fishing and Aquaculture. These committees can address a) closed access fisheries, in recovery or initial development, and b) fisheries of benthic resources. They are in charge of designing, evaluating and implementing management plans, which are created to establish norms and actions to allow fisheries management within the framework of sustainability and conservation of resources. In 2019, there are 33 operational Management Committees that have approved 18 management plans.

Chile and Mexico also have fisheries management plans which are subject to periodic evaluations to assess whether objectives were met or not. Peru still requires strengthening its decision-making processes and explicitly defining management objectives and goals. Together with information from the Fisheries Management Regulations, these should be integrated into public management plans (Heck, 2015).

It is important to note that the three countries are strengthening their fisheries management institutions as well as their supervision and sanction systems. For example, the Regulation on the Organization and Functions of the Ministry of Production in Peru was modified in order to facilitate, among others, the execution of functions by the Ministry regarding policy development and planning, as well as supervision, inspection and sanction imposition. In Chile, the new Fishing Law strengthens the research role of the Institution for the Promotion of Fishing and inspection responsibilities under the National Service for Fishing and Aquaculture.

In terms of transparency, Peru is undertaking considerable efforts. For instance, it is the first country in the region – and second in the world – to allow the publication of all information gathered by its satellite tracking system on the portal of Global Fishing Watch.[12] This initiative makes information available on an open access virtual platform regarding the areas where national and foreign industrial fleets are operating. This allows citizens, governments, companies and non-governmental organizations to monitor the conduct of fishing vessels in real time, thus strengthening national monitoring systems to combat illegal fishing.

In contrast, this is an aspect which requires considerable strengthening in Mexico. Fishing vessels are required by law to submit information on their fishing voyages. However, most vessels don't and sanctions are not applied for this violation (Alfaro and Quintero, 2014).

Chile, Mexico and Peru face a common challenge: how to endow their research institutions with the necessary resources and autonomy. To this effect, their plans and research budgets must be aligned with each other. This will guarantee financial security and prevent disruptions in the research efforts and/or changes in research priorities due to personnel rotation or variations in funding priorities by external donors.

Fishery governance systems and the institutions involved must be capable of addressing the challenges and opportunities posed by climate change. Although climate change has not been discussed previously in this chapter, it is important to note that it is a crosscutting dimension. Climate change is affecting the structure and functions of marine ecosystems, generating variations in their global productivity and stability (Cheung *et al.*, 2009; Cheung *et al.*, 2010; Sumaila *et al.*, 2011). Accordingly, it is acknowledged that the sustainable management of fisheries, including recovery of overexploited stocks, is the best and most effective climate change adaptation measure (Sumaila *et al.*, 2012). Most critically, fisheries management systems must have an ecosystem approach perspective, which is articulated with effective and dynamic governance systems (i.e. adaptative and proactive) (Pauly *et al.*, 2002; Allison *et al.*, 2009; Sumaila *et al.*, 2012; Miller *et al.*, 2013; Bennett and Satterfield, 2018).

Additionally, governance systems must encourage the implementation and maintenance of communication channels among different users of the marine ecosystems and their managers. This is necessary, as industrial fishing does not take place in isolation from other activities (i.e. artisanal fishing, tourism and maritime transport) and/or social interests (i.e. biodiversity conservation, food security and job creation). Therefore, governance should be contextualized within the "blue economy" concept (UNEP, 2015). This means that different users of marine areas work together and agree on strategies to encourage the improvement of their quality of life and local social welfare by planning the use and zoning of marine ecosystems.

Finally, it is important to highlight the work undertaken by different Peruvian actors in order to strengthen governance with regards to anchovy industrial fishing. This type of fishing relies on a tracking system for landings financed by companies and executed by certified supervisory institutions together with officials from the Ministry of Production and the Institute of the Sea of Peru (De la Puente *et al.*, 2011). This system is a kind of census and covers all landing points and ports for industrial vessels (Heck, 2015). It is useful both for monitoring catches and quantifying juvenile landings which have a negative impact on the sector's revenues (Salvatteci and Mendo, 2005). The framework that regulates fishing seasons includes provisions regarding temporary closure of fishing zones when juvenile landings, in a determined port, exceed 10 percent of daily landings. However, this is a reactive measure and closures are few and inappropriate as vessels will tend to discard juveniles *before* landings and reporting (De la Puente *et al.*, 2011). Nonetheless, industrial vessels of some companies did report to their control centers the areas where juveniles were being captured. This information was then forwarded to all the vessels of their fleet in order to avoid fishing in these areas. This approach to juvenile catch was presented to the Ministry of Production by members of the National Fishery Union and given its effectiveness, led to the mandatory and preventive closure of various fishing zones, which overall reduced bycatch of juveniles.

This is an example of good communication among actors of the sector and an opening to learning for the authorities. This collaboration makes it possible to adapt and strengthen management mechanisms, contributing to the sustainable use of natural resources.

Conclusions

Industrial fishing has evolved into an extremely efficient extractive activity, both in Latin America and the rest of the world. It is essential therefore that management and planning mechanisms to safeguard the sustainable use of marine resources and ecosystems are developed and implemented. Throughout this chapter the challenges faced by Chile, Mexico and Peru in the management of their industrial fisheries has been presented.

First, sustainable management of target resources has to be sought and achieved. To this effect, it is key that the exploitation rates for resources are determined based on biological reference points that incorporate the variability and dynamism of ecosystems. The example of the determination of north-center anchovy stock quotas in Peru is a good example of successful management. Likewise, the need for specific and measurable objectives for fisheries management is key. In this respect, both Mexico and Chile, unlike Peru, rely on fisheries management plans with concrete and measurable objectives and goals.

Second, the ecosystem approach needs to be integrated into fisheries management. The decision-making process should be based on quantitative analysis of the combined impact by different fisheries on the various components of the ecosystem, incorporating the best science possible and a precautionary approach in the face of uncertainty. Chile, Mexico and Peru still need to improve public funding for national research institutions whose results feed management decisions and improve regulatory frameworks. The prevention and mitigation of bycatch and discard also need to be prioritized. In this regard, Chile is a good example considering it has a plan in place for the reduction of discard and has identified vulnerable marine ecosystems where trawl fishing is banned while Peru has also taken a significant step with the establishment of preventive closures in zones where juvenile bycatch takes place.

Third, good fisheries governance is key. This must be effective, equitable, proactive, robust, dynamic and adaptative. Each country has made important progress in specific areas of governance. Nonetheless, there is in general limited progress in overall sustainability through appropriate institutional and regulatory frameworks, supervision and sanction systems, citizen participation mechanisms, research and monitoring and evaluation.

It is important to take into account that industrial fishing does not take place in an isolated context – it relates to other economic activities and/or interests. Therefore, to improve the quality of life of the people, prevent conflicts, address climate change and minimize the overexploitation of live marine resources, the "blue economy" notion offers a good framework under which to contextualize development of coastal and marine areas. An initial step in this direction is to establish communication channels among different groups of users of these ecosystems and managers. Good communication generates confidence and respect among actors, promotes joint work and strengthens governance systems.

Lastly, industrial fishing is an extremely important economic activity for Chile, Mexico and Peru. Regardless of the present state of fisheries and resources and the uncertainties of climate change and its effects, implementation of sustainable management for resources under an ecosystem approach, would allow continued recovery and overall development. Industrial fishing continues having the potential to increase contributions to employment, earnings and food security in the near future.

Notes

1 Fishery Law (Law No. 19,713), which entered into force in 1991.
2 New Fisheries Law of Chile (Law No. 20,657), which entered into force in 2013, replacing Law No. 19,713.
3 General Law of Fishing of Peru (Law Decree No. 25,977), which entered into force in 1992.
4 Regulation of the General Law of Fishing of Perú, approved by Supreme Decree No. 012-2001-PE, which entered into force in 2001.
5 Modification to the Regulation of Peru's General Law of Fishing, approved by Supreme Decree No. 006-2018-PRODUCE which entered into force in 2018.
6 Fisheries Management Regulation for Peruvian Hake (*Merluccius gayi peruanus*), approved by Supreme Decree No. 016-2003-PRODUCE, which entered into force in 2003.
7 Law on Maximum Catch Limits per Vessel for Industrial Anchovy Fishing (*Engraulis ringens*), approved by Legislative Decree No. 1,084, which entered into force in 2009.
8 Regulation on the Organization and Function of PRODUCE, approved by Supreme Decree No. 002-2017-PRODUCE, which entered into force in 2017.
9 Procedure to preventively suspend activities in areas where there is anchovy in sizes smaller than permitted, approved by Directorial Resolution No. 012-2014-PRODUCE/DGSF, in force since 2014.

10 EcoBase official portal, digital repository of complex ecological models. Available at, http://sirs.agrocampus-ouest.fr/EcoBase/#discoverytools.
11 Regulation for control and sanctions regarding fishing and aquaculture activities in Peru, approved by Supreme Decree No. 017-2017-PRODUCE, which entered into force in 2017.
12 Global Fishing Watch, see, https://globalfishingwatch.org.

References

Alfaro, S. B., and Quintero, M. L. (2014), Sector Pesquero-acuícola en México y Chile: Estudio de Caso Comparativo para Reflexionar Respecto de su Internalización. *AgroSur* 42, 31–46.

Allison, E. H., Perry, A. L., Badjeck, M.-C., Neil Adger, W., Brown, K., Conway, D., *et al.* (2009), Vulnerability of National Economies to the Impacts of Climate Change on Fisheries. *Fish and Fisheries*, 10(2), 173–196.

Anderson, L. G., and Seijo, J. C. (2010), *Bioeconomics of Fisheries Management.* Wiley-Blackwell. Iowa, USA.

Arias-Schreiber, M. (2012), The Evolution of Legal Instruments and the Sustainability of the Peruvian Anchovy Fishery. *Marine Policy*, 36(1), 78–89.

Arreguín-Sánchez, F., and Arcos-Huitrón, E. (2011), La Pesca en México: Estado de la Explotación y Uso de los Ecosistemas. *Hidrobiológica* 21, 431–462.

Bennett, N. J., and Satterfield, T. (2018), Environmental Governance: A Practical Framework to Guide Design, Evaluation, and Analysis. *Conservation Letters*, 11(6), e12600–13.

Biblioteca del Congreso Nacional de Chile – BCN (2014), *La Legislación Pesquera y las Modificaciones Producto de la Ley Número 20.657.* BCN Informe.

Caddy, J. F. (1999), Fisheries Management in the Twenty-First Century: Will New Paradigms Apply? *Reviews in Fish Biology and Fisheries*, 9(1), 1–43.

Cashion, T., Al-Abdulrazzak, D., Belhabib, D., Derrick, B., Divovich, E., Moutopoulos, D. K., *et al.* (2018a), Reconstructing Global Marine Fishing Gear Use: Catches and Landed Values by Gear Type and Sector. *Fisheries Research*, 206, 57–64.

Cashion, T., De la Puente, S., Belhabib, D., Pauly, D., Zeller, D.,and Sumaila, U. R. (2018b), Establishing Company Level Fishing Revenue and Profit Losses from Fisheries: A Bottom-up Approach. *PLoS ONE*, 13(11), e0207768–20.

Cashion, T., Le Manach, F., Zeller, D., and Pauly, D. (2017), Most Fish Destined for Fishmeal Production are Food-Grade Fish. *Fish and Fisheries*, 18(5), 837–844.

Castilla, J. C. (2010), Fisheries in Chile: Small Pelagics, Management, Rights and Sea Zoning. *Bulletin of Marine Science*, 86(2), 221–234.

Cheung, W. W. L., Lam, V. W. Y., Sarmiento, J. L., Kearney, K., Watson, R., and Pauly, D. (2009). Projecting Global Marine Biodiversity Impacts under Climate Change Scenarios. *Fish and Fisheries*, 10(3), 235–251.

Cheung, W. W. L., Lam, V. W. Y., Sarmiento, J. L., Kearney, K., Watson, R., Zeller, D., and Pauly, D. (2010). Large-scale redistribution of maximum fisheries catch potential in the global ocean under climate change. *Global Change Biology*, 16(1), 24–35.

Christensen, V., De la Puente, S., Sueiro, J. C., Steenbeek, J., and Majluf, P. (2014a), Valuing Seafood: The Peruvian Fisheries Sector. *Marine Policy*, 44, 302–311.

Christensen, V., Coll, M., Piroddi, C., Steenbeek, J., Buszowski, J., and Pauly,

D. (2014b), A Century of Fish Biomass Decline in the Ocean. *Marine Ecology Progress Series*, 512, 155–166.

Cisneros-Montemayor, A. M., Cisneros-Mata, M. A., Harper, S., and Pauly, D. (2013), Extent and Implications of IUU Catch in Mexico's Marine Fisheries. *Marine Policy*, 39, 283–288.

Cisneros-Montemayor, A. M., Cisneros-Mata, M. A., Harper, S., and Pauly, D. (2015), *Unreported Marine Fisheries Catch in Mexico, 1950–2010*. Fisheries Centre Working Paper #2015–22, University of British Columbia, Vancouver.

Cisneros-Montemayor, A. M., Sanjurjo, E., Munro, G. R., Hernández–Trejo, V., and Rashid Sumaila, U. (2016), Strategies and Rationale for Fishery Subsidy Reform. *Marine Policy*, 69, 229–236.

Cochrane, K. L. (ed.) (2002), *A Fishery Manager's Guidebook. Management Measures and Their Application.* FAO Fisheries Technical Paper. No. 424. Rome, FAO.

Cochrane, K. L. (2005), Guía del Administrador Pesquero: Medidas de Ordenación y su Aplicación. FAO Documento Técnico de Pesca No. 424. Rome.

Daly, H. E. (2005), Economics in a Full World. *Scientific American*, 293(3), 100–107.

De la Puente, O., Sueiro, J. C., Heck, C., Soldi, G., and De la Puente, S. (2011), *La Pesquería Peruana de Anchoveta: Evaluación de los Sistemas de Gestión Pesquera en el Marco de la Certificación a Cargo del Marine Stewardship Council.* Centro para la Sostenibilidad Ambiental de la Universidad Peruana Cayetano Heredia. Lima, Peru.

EDF. (2018), *Map of Sustainable Fisheries. Fishery Solutions Center.* Available at http://fisherysolutionscenter.edf.org/map. Accessed December 10, 2018.

Ferguson-Cradler, G. (2018), Fisheries' Collapse and the Making of a Global Event, 1950s–1970s. *Journal of Global History*, 13(3), 399–424.

Graham, J., Amos, B., Plumptre, T. (2003), *Principles for Good Governance in the 21st Century.* Policy Brief No. 15. Institute on Governance. Ottawa, Canada.

Halpern, B. S., Walbridge, S., Selkoe, K. A., Kappel, C. V., Micheli, F., D'Agrosa, C., et al. (2008), A Global Map of Human Impact on Marine Ecosystems. *Science*, 319(5865), 948–952.

Heck, C. (2015), *Hacia un Manejo con Enfoque Ecosistémico de la Pesquería Peruana de Achoveta. Análisis del Marco Legal y Administrativo para Reformar el Manejo de la Pesquería Peruana de Anchoveta.* SPDA/Earthjustice/AIDA. Lima, Peru.

Instituto Nacional de Pesca – INAPESCA (2018), Carta Nacional Pesquera 2017. Diario Oficial de la Federación, México.

Larkin, P. A. (1977), An Epitaph for the Concept of Maximum Sustained Yield. *Transactions of the American Fisheries Society*, 106(1), 1–11.

Mangin, T., Costello, C., Anderson, J., Arnason, R., Elliott, M., Gaines, S. D., et al. (2018), Are Fishery Management Upgrades Worth the Cost? *PLoS ONE*, 13(9), e0204258–24.

Mendo, J., and Wosnitza-Mendo, C. (2014), *Reconstruction of Total Marine Fisheries Catch for Peru: 1950–2010.* Fisheries Centre Working Paper #2014–21.

Miller, K. A., Munro, G. R., Sumaila, U. R., and Cheung, W. W. L. (2013). Governing Marine Fisheries in a Changing Climate: A Game-Theoretic Perspective. *Canadian Journal of Agricultural Economics/Revue Canadienne D'agroeconomie*, 61(2), 309–334.

Miller, M. (2014), *¿Quién Debe Pagar los Costos de Administración de un Sistema de Cuotas Individuales Transferibles?: Una Aplicación Basada en Simulaciones Numéricas a la Pesquería de Sardina Común y Anchoveta de la Zona Centro-sur de Chile.* Universidad de Concepción, Chile.

Myers, R. A., and Worm, B. (2003), Rapid Worldwide Depletion of Predatory Fish Communities. *Nature*, 423(6937), 280–283.

O'Keefe, C. E., Cadrin, S. X., and Stokesbury, K. D. E. (2014), Evaluating Effectiveness of Time/Area Closures, Quotas/Caps, and Fleet Communications to Reduce Fisheries Bycatch. *ICES Journal of Marine Science*, 71(5), 1286–1297.

Oliveros-Ramos, R., and Díaz Acuña, E. (2015), *Estimación de la Captura Total Permisible del Stock Norte Centro de la Anchoveta Peruana.* Protocolo del Instituto del Mar del Perú. IMP-DGIRP/AFDPERP. Callao, Peru.

Olsen, E., Aanes, S., Mehl, S., Holst, J. C., Aglen, A., and Gjøsæter, H. (2010), Cod, Haddock, Saithe, Herring, and Capelin in the Barents Sea and Adjacent Waters: A Review of the Biological Value of the Area. *ICES Journal of Marine Science*, 67: 87–101.

Panayotou, T. (1983), *Conceptos de Ordenación para las Pesquerías en Pequeña Escala: Aspectos Económicos y Sociales.* FAO Documento Técnico de Pesca No. 228. Rome.

Pauly, D. (2009), Beyond Duplicity and Ignorance in Global Fisheries. *Scientia Marina*, 73(2), 215–224.

Pauly, D., and Zeller, D. (2016), Catch Reconstructions Reveal that Global Marine Fisheries Catches are Higher than Reported and Declining. *Nature Communications*, 7, 10244.

Pauly, D., Christensen, V., Guénette, S., Pitcher, T. J., Sumaila, U. R., Walters, C. J., *et al.* (2002). Towards sustainability in world fisheries. *Nature*, 418(6898), 689–695.

Pikitch, E. K., Rountos, K. J., Essington, T. E., Santora, C., Pauly, D., Watson, R., *et al.* (2014), The Global Contribution of Forage Fish to Marine Fisheries and Ecosystems. *Fish and Fisheries*, 15(1), 43–64.

Rochet, M.-J., Catchpole, T., and Cadrin, S. (2014), Bycatch and Discards: from Improved Knowledge to Mitigation Programmes. *ICES Journal of Marine Science*, 71(5), 1216–1218.

Salvatteci, R., and Mendo, J. (2005), Estimación de las Pérdidas Bio-económicas Causadas por la Captura de Juveniles de Anchoveta (*Engraulis ringens*, J.) en la Costa Peruana. *Ecología Aplicada*, 4(1–2), 113–120.

Sapag, C., Gorny, M., and Van der Meer, L. (2016), *Propuesta para la Eliminación Progresiva de la Pesca de Arrastre en Chile.* Oceana Chile. Santiago, Chile.

Seitz, R. D., Wennhage, H., Bergström, U., Lipcius, R. N., and Ysebaert, T. (2014), Ecological Value of Coastal Habitats for Commercially and Ecologically Important Species. *ICES Journal of Marine Science*, 71(3), 648–665.

SUBPESCA (2017), *SUBPESCA establece los Primeros planes de Reducción del Descarte y la Captura de Pesca Incidental.* Available at https://goo.gl/VEKSjA. Accessed November 30, 2018.

SUBPESCA (2018), *Estado de Situación de las Principales Pesquerías Chilenas*, Año 2017. Subsecretaría de Pesca y Acuicultura. Santiago, Chile.

Sumaila, U. R., Cheung, W. W. L., Lam, V. W. Y., Pauly, D., and Herrick, S. (2011), Climate Change Impacts on the Biophysics and Economics of World Fisheries. *Nature Climate Change*, 1(9), 449–456.

Sumaila, U. R., Cheung, W., Dyck, A., Gueye, K., Huang, L., Lam, V., *et al.* (2012), Benefits of Rebuilding Global Marine Fisheries Outweigh Costs. *PLoS ONE*, 7(7), e40542–12.

Teh, L. C. L., and Pauly, D. (2018), Who Brings in the Fish? The Relative Contribution of Small-Scale and Industrial Fisheries to Food Security in Southeast Asia. *Frontiers in Marine Science*, 5, 541–9.

Tveteras, S., Paredes, C. E., and Peña-Torres, J. (2011), Individual Vessel Quotas in Peru: Stopping the Race for Anchovies. *Marine Resource Economics*, 26(3), 225–232.

UNEP. (2015), *Blue Economy: Sharing Success Stories to Inspire Change*. UNEP Regional Seas Report and Studies No. 195. Geneva, Switzerland. 44p.

van der Meer, L., Arancibia, H., Zylich, K. and Zeller, D. (2015), *Reconstruction of Total Marine Fisheries Catches for Mainland Chile (1950–2010)*. Fisheries Centre Working Paper #2015–91, University of British Columbia, Vancouver.

Walters, C. J., and Martell, S. J. D. (2004), *Fisheries Ecology and Management*. Princeton University Press. New Jersey, USA.

Worm, B., Barbier, E. B., Beaumont, N., Duffy, J. E., Folke, C., Halpern, B. S., *et al.* (2006), Impacts of Biodiversity Loss on Ocean Ecosystem Services. *Science*, 314(5800), 787–790.

3   Marine bioprospecting

Liliana Pardo López

Introduction

The oceans cover more than 70 percent of the Earth's surface. Over 30 of the 36 known animal phyla inhabit the oceans and sea. It is estimated that the total number of marine species is 1 million, together with hundreds of millions of microbial species. Marine genetic resources are becoming an important input to biotechnology and bioprospecting, given marine organisms can survive environments with high atmospheric pressures, extremely high and low temperatures, low oxygen levels, high salt concentrations and even lack of light. Their metabolisms have adapted to contend with extreme environmental pressures. As a result, expectations for finding novel and useful industrial organisms or natural products derived thereof, are very high and on the rise (Skropeta and Wei, 2014).

This chapter will reflect briefly on the international environmental policies and laws related to biodiversity and marine bioprospecting in particular. Examples of marine bioprospecting will be described to provide an overall picture of advances in the field. It will also address issues of national and international jurisdiction or the lack thereof, regarding marine genetic resources (MGR). Although there has been a recent increase in the number of papers and publications regarding marine bioprospecting, the issue is relatively new in Latin America. Its technology, financial and knowledge-intensive features have made marine bioprospecting a field closed almost exclusively for developed countries. Understanding some of the social and economic implications of this activity will hopefully offer decision makers, scholars and interested actors an avenue for further and more detailed analysis of marine bioprospecting. A section on conclusions and challenges closes this overview and reflection of a potentially useful and lucrative activity, with considerable implications on sustainable development.

Conceptual issues: genetic resources, biotechnology and bioprospecting

What are genetic resources?

The most popular and widely accepted legal definition for genetic resources in that of the Convention on Biological Diversity (CBD) of 1992, which defines "genetic resources" as "any material of plant, animal, microbial or other origin containing functional units of heredity." "Biological resources" on the other hand, include "genetic resources, organisms or parts thereof, populations, or any other biotic component of ecosystems with actual or potential use or value for humanity."

What is biotechnology?

The CBD also defines biotechnology as "any technological application that uses biological systems, living organisms, or derivatives thereof, to make or modify products or processes for specific use." As with any other technological tool, biotechnology should be deployed to solve problems in the environmental, health, food and agricultural sectors in a respectful and sustainable manner.

It is difficult to specify the exact birth date of biotechnology given the long history of manipulation of biological systems. Examples of biotechnology include the development of the first smallpox vaccine by Alexander Fleming in the early 20th century and the creation of the first transgenic organisms that allowed the synthesis of a human hormone in bacteria in 1979. The discovery in 1953 of the double helix of deoxyribonucleic acid (DNA) by James Watson and Francis Crick, paved the way for genetic engineering and *modern* biotechnology which includes fields such as molecular biology, microbiology, genomics, gene editing and biochemical engineering (Bolivar-Zapata, 2017).[1]

Marine microorganisms (bacteria, viruses, phytoplankton, zooplankton), as well as their derivatives (DNA, enzymes, secondary metabolites) are a valuable source of inputs for modern biotechnology. In the past, access to marine microorganisms or genetic material was possible mostly through laboratory cultures. However, it is estimated that researchers have only managed to cultivate less than 1 percent of all bacteria living on the planet. Recently, independent cultivation techniques have been developed, allowing for genetic material from microorganisms living in the sea to be obtained, including from abyssal zones.[2] Metagenomics is one such technique which allows access to previously unexplored resources.[3]

What is bioprospecting?

Bioprospecting can be defined as the systematic search for new sources of useful and potentially commercially valuable chemical compounds, genes, proteins, secondary metabolites, microorganisms, etc., from nature and biodiversity. This search demands responsible management of resources (e.g. plants, animals, fungi and microorganisms) and environmentally responsible conducts and practices. Marine bioprospecting, as the name suggests, utilizes marine compounds and organisms from a variety of ecosystems such as the seabed, hydrothermal vents, seamounts and coral reefs.

For bioprospecting to take place, research must first be undertaken on new organisms and biological sources which may provide useful compounds.

A biotechnology intermediate step then takes place and, finally, a move towards commercializing the possible product/process. Coordinated work among scientists, business and the government is essential for these steps to happen fluidly. Strategic alliances should be established to formulate bioprospecting plans which contribute to national development. Ethical considerations and recognition of the value of nature and its contribution to sustainable livelihoods should be part of these strategies and plans (Gómez-Madrigal *et al.*, 2014).

International environmental law: relevant milestones

United Nations Convention on the Law of the Sea (UNCLOS)

The first United Nations Conference on the Law of the Sea was held in 1956. In 1994 the United Nations Convention on the Law of the Sea (UNCLOS)[4] was adopted and is now considered analogous to a "Constitution for the Oceans" which is binding to 168 parties. UNCLOS recognizes the right of states to establish a territorial sea (12 nautical miles), an adjacent area (12 nautical miles) and an Exclusive Economic Zone (EEZ) (200 nautical miles). Within the EEZ each state has sovereign rights for exploration and exploitation purposes, conservation and management of natural resources, as well as for marine research and protection and preservation of the marine environment. Although UNCLOS did not originally refer to biological or genetic resources, there is at present a move within the agreement to develop a specific legal regime which addresses genetic resources and related digital sequence genetic information and aligns its principles with the CBD and the Nagoya Protocol on Access to Genetic Resources and the Fair and Equitable Sharing Arising from their Utilization (2010).

Convention on Biological Diversity (CBD)

The CBD was signed on June 5, 1992 at the United Nations Conference on Environment and Development in Rio de Janeiro (UNCED or "Earth Summit") and entered into force on December 29, 1993, after its ratification by 193 Parties.[5] It has three main goals including: the conservation of biological diversity; the sustainable use of its components; and the fair and equitable sharing of benefits arising from the use of genetic resources (Glowka *et al.*, 1994). The CBD covers biotechnology extensively, recognizing its critical potential in the sustainable use of genetic resources – both marine and terrestrial. Articles 15, 16 and 19 specifically address issues related to the rules of access to genetic resources and benefit sharing, access to and transfer of biotechnology and its handling, and the distribution of its benefits, respectively. It is within these principles that bioprospecting develops. The CBD specifies that ecosystems, species and genetic resources must be utilized for the benefit of humans, without undermining biological diversity. Its binding, albeit often qualified, provisions also address principles of cooperation, awareness raising, *in situ* and *ex situ* conservation, sustainable use, among others.

Cartagena Protocol on Biosafety

The UNCED Declaration on Environment and Development, refers to the precautionary approach in the transfer and manipulation of living modified organisms (i.e. genetically modified organisms produced through modern biotechnology).[6] This specific principle and the CBD references to the safe use and handling of biotechnological products, paved the way for the Cartagena Protocol on Biosafety (2000).[7] The Cartagena Protocol contributes to ensure an adequate protection in the transfer, handling and use of living modified organisms resulting from modern biotechnology that may have adverse effects on the conservation and sustainable use of biological diversity, also taking into account risks to human health. Its emphasis is international or transboundary movements of these organisms.

Mexico was one of 170 countries to sign the Cartagena Protocol and according to its monitoring and reporting obligations, the Secretariat for the Environment and Natural Resources, the Secretariat for Health,[8] the Secretariat for Cattle, Rural Development, Fishing and Food,[9] the National Commission for the Knowledge and Use of Biodiversity and Inter-Secretarial Commission on Biosafety of Genetically Modified Organisms have submitted periodic reports on monitoring and implementation of the Protocol to the CBD's Secretariat. The Biosafety Law on Genetically Modified Organisms (2005) and the regulation on the functions of the Inter-Secretarial Commission on Biosafety of Genetically Modified Organisms are two reflections of advances made in this field.[10] Chile and Peru have also signed the Cartagena Protocol[11] and together with Colombia have been working on a multinational project – OEA-CONICYT – focused on the evaluation of policies and national regulatory systems for biosafety at the national level to facilitate the implementation of the Protocol.

Nagoya Protocol on Access to Genetic Resources

The Nagoya Protocol on Access to Genetic Resources and the Fair and Equitable Sharing of Benefits Arising from their Utilization was adopted in 2010 in Nagoya, Japan, after several years of arduous negotiations.[12] It includes a series of provisions to promote benefit sharing through, for example, transfer of biotechnology, streamlining access and benefit-sharing procedures, developing monitoring mechanisms and check points to very movements and uses of genetic resources, etc. One key goal of the Nagoya Protocol is to create legal certainty for and transparency between users and providers of genetic resources and related traditional knowledge of indigenous peoples.

Mexico ratified the Nagoya Protocol on May 16, 2012 and it is now in the process of being implemented through a Global Environmental Facility and United Nations Development Program project. The key principle underlying the implementation process is the recognition that states have the sovereign right to exploit their own natural resources pursuant to environmental safeguards and to ensure that these activities do not undermine other states or zones situated outside national jurisdiction.[13] The project seeks to contribute to the construction of sensible laws and regulations to facilitate and control access to and use of genetic resources. The project is conducted by the Secretariat for the Environment and Natural Resources and involves key actors related to research in genetic resources, including universities, civil society and indigenous and local communities. To consolidate and ensure appropriate participation, an interagency working group has been created and includes the Secretariat for the Environment and Natural Resources, the Secretariat for Cattle, Rural Development, Fishing and Food, the Secretariat for Foreign Relations, the National Commission for the Development of Indigenous Peoples and the Secretariat for Economy.

Regulating marine genetic resources (MGR) within or beyond national jurisdictions

Marine genetic resources include plants, animals and microorganisms, either complete organisms or groups of organisms such as plankton or samples partially contained in water and sediment.[14] Genomics data covers deoxyribonucleic acid (DNA) sequences obtained through different methodologies such as Sanger sequencing or new massive sequencing techniques (illumine, PacBio, Oxford Nanopore). Marine samples are used to classify organisms or microorganisms for 16S or 18S ribosomal gene sequencing. Metagenomics has revolutionized the way in which marine microorganisms are classified. Beginning in the 21st century important contributions have been made in this field.[15] For instance, the ocean exploration genome project – Global Ocean Sampling Expedition (GOS) – conducted by the Craig Venter Institute, whose goal is to assess genetic diversity in marine microbial communities, has visited 23 countries spanning four continents, reporting millions of new genes and nearly 1,000 genomes for uncultivated microbial lineages, as well as a wider understanding of marine microbiology. In Mexico, the Consortium for the Research of the Gulf of Mexico (2015–2020)[16] conducts a similar project of oceanographic expeditions seeking to obtain a regional baseline for marine genetic resources, using metagenomics techniques (Escobar-Zepeda *et al.*, 2018; Godoy-Lozano *et al.*, 2018). Hand-in-hand with the development of genomics and metagenomics, bioinformatics is having a key role in analyzing and classifying large amounts of data generated on a daily basis. However, the oceanographic community is not immune to problems with regards to unloading, classifying and analyzing metadata about genome sequences or associated to them. This community faces storage limitations for samples (genetic resources) obtained from the sea. Clear and definitive international and national regulations are nonexistent to date, therefore each university, gene bank or company stores genetic resources according to different protocols and only a few actually report to countries in whose jurisdictions these marine genetic resources were obtained.

Deep-sea marine areas, especially Areas Beyond National Jurisdiction (ABNJ), are particularly challenging from an institutional and policy perspective.

These areas comprise 64 percent of the ocean's surface around the world and 47 percent of the earth's surface (Costello *et al.*, 2017). There are no international instruments or national laws to govern collection and storage of samples and related data, let alone addressing property and control dimensions, particularly in the context of increasing enclosures through patents over innovations based on these resources. It is important to take into account that the majority of species found in these areas have yet to be taxonomically described and genomically deciphered. Transnational corporations and northern hemisphere research institutions with technological capacities have a large advantage in exploring and making use of marine genetic resources.

Some efforts have been undertaken recently to compile international registers of sample collections including through the Global Genome Biodiversity Network (GGBN) that saves genomic collections for research purposes[17] and the Global Registry of Biodiversity Repositories (Schindel *et al.*, 2016).[18] These international initiatives are supported as a means to improve storage capacities and create standardized principles and best practices regarding collecting, storing and analyzing marine genetic resources. Open data repositories can improve access to data and information and strengthen opportunities for collaboration.

Whilst bioprospecting is a technically complex and challenging endeavor, sensible policies and legal frameworks to address it are still in the making. More than two decades since the CBD was adopted, countries and the international community are still grappling with how to develop appropriate rules, even in the light of advancements with the FAO International Treaty and Nagoya Protocol (Wright *et al.*, 2016). One looming question is how to regulate marine genetic resources and digital sequence information derived thereof. Quite a few people advocate that marine genetic resources, particularly beyond national jurisdiction, should be considered part of the world heritage and freely accessible for research and development, in practice moving away from the sovereignty principles in the CBD and Nagoya Protocol. However, especially developing countries, some consider this may be unfair given capacities to access and use these resources are concentrated in very few countries and institutions. They also call for bioprospecting or marine genetic resources to be aligned with the CBD and Nagoya Protocol benefit sharing principles.

The challenge for developing countries to participate in world biotechnology markets

Governments around the world are seeking opportunities for sustainable economic growth. Most acknowledge that development and success cannot place an additional cost on the planet's already endangered ecosystems. Many countries see in the "bioeconomy" – parts of the economy that use renewable biological resources from land and sea, to produce food, materials and energy – a key conceptual framework under which to plan their development. For instance, it is estimated that the global market for marine biotechnology has the potential to reach US$4.8 billion by 2020 and US $6.4 billion by 2025. The identification of new applications for enzymes derived from the sea and uses of bacteria, marine algae and microalgae in the production of biofuels can become key drivers for market growth and a dimension of development (Hurst *et al.*, 2016).

Unfortunately for the interests of developing countries, access to and use of marine resources, particularly outside their jurisdiction and in the deep-sea bed, are disproportionately concentrated in companies, research centers and transnational corporations with the financial clout and technical capacities to undertake these high-risk projects. The literature reviewed shows that 84 percent of patents derived from marine products are owned by a small group of companies. The chemical company BASF, headquartered in Germany, is the largest chemical producer in the world. It owns 47 percent of all marine sequences, including patented gene sequences. In 2017, the company surpassed US$79 billion in registered sales, with a network of 633 subsidiaries worldwide. Germany (49 percent), United States (13 percent) and Japan (12 percent) are the three countries with most patent applications – and granted patents – associated to innovations derived from marine genetic resources (Blasiak *et al.*, 2018). Most patents are under company control, with universities in contrast holding only 12 percent of registered patents. These figures open at least two important considerations. First, bioprospecting is not only technically complex, but it often also involves very complex relationships between multiple actors; second, patents over biological materials immediately call into the analysis whether samples and genetic resources were accessed in accordance with the CBD and Nagoya Protocol principles on benefit sharing and whether these have been fairly and equitably shared.

Mexico, Chile and Peru have very rich marine environments and some of the most important fisheries in the world. Except for a few well documented cases, there is little information available as to exactly what is happening in terms of bioprospecting of marine resources. For instance, in Mexico, the technology company Nova Proteins produces shark antibodies to generate various medical applications.[19] In Peru, Peruvian Seaweed (PSW) is a successful technology company discreetly producing biotechnological products based on marine resources, especially algae. Many of its products are already protected by trademark and patents associated to their products and processes – which is quite exceptional for most developing countries.[20] However, Chile, Mexico and Peru, as do most developing and biodiversity endowed countries, act as providers of biological and genetic resources which then feed into technological processes in industrialized nations. Nuances are warranted with regards to this scenario, but it describes quite correctly a historic pattern.

Institutional, policy and legal challenges for marine bioprospecting in Mexico

Development and implementation of legal measures in Mexico regarding access and benefit sharing in accordance with the Nagoya Protocol, has been slow and there are numerous challenges to be overcome. A national plan or bioprospecting strategy demands government leadership and involvement of a wide range of institutions from civil society, academia and universities, private companies and state agencies.

Some countries in Latin America have developed a series of strategies and planning instruments which specifically address bioprospecting. For example, Colombia has approved a National Marine Bioprospecting Plan (Melgarejo *et al.*, 2002)[21]; Peru has in turn approved a Biodiversity Valorization Program which includes specific lines of work and funding for the implementation of bioprospecting activities.[22] General biodiversity strategies in Chile, Mexico and Peru all include specific mandates and describe actions to undertake research and development on genetic resources, with due consideration of the CBD and Nagoya Protocol principles.

Mexico has over the past few years developed a considerable and strong capacity in biotechnology shared among universities, private companies and research institutions. The recent ABS project sponsored by the Global Environmental Facility in Mexico, offers a formidable opportunity to develop a strategy and plan for marine bioprospecting in particular.

A national baseline informed by the business sector, academia and universities, indigenous and local communities and governmental organizations would be a first step to understand the types of marine research being undertaken throughout Mexican territorial seas and support further decisions. The Secretariat of Finance and Public Credit and the National Evaluation Committee define the priorities for GEF and other projects and approve their development and implementation.

This initial step, integrating marine bioprospecting into the national agenda, could be followed by an effort to identify comparative marine bioprospecting cases around the world and best practices in regulatory implementation and institutional development. This could also lead to an analysis of the implications of intellectual property and its effects on stimulating research and development in marine genetic resources. Ultimately, understanding value chains and markets also requires multisectoral and interdisciplinary analysis to identify potential commercial and industrial opportunities.

There is no way bioprospecting and biotechnology development can take place in isolation. The breath of collaboration to strengthen and consolidate successful research and development endeavors in marine genetic resources is very considerable. National and international consortia are common and should be encouraged. Very importantly, science and technology funding agencies such as the National Council for Science and Technology in Mexico or the National Technology Council in Peru, need to increase investments in research and development in marine bioprospecting in particular. Mexico should include a marine bioprospecting section in its recent National Plan for Science and Technology 2018–2024.[23]

Brief bioprospecting cases

It is difficult to estimate the economic value of the different forms of benefits derived from marine bioprospecting activities. There are, however, some useful indicators. Hundreds of patents have been granted during the last 30 years over marine resources derived innovations; as part of these patents, 862 different marine species have been identified; and a total of 12,998 genetic sequences are associated to those patents. Almost 73 percent of patents relate to microorganisms; fish and mollusks represent 16 percent and 3 percent respectively. Approximately 11 percent of patents derive from species living in deep seas and hydrothermal vents, many found beyond national jurisdictions (Blasiak *et al.*, 2018).

There are a group of well-known and documented examples of successful or, at least, persistent bioprospecting efforts around the world. One of the best-known cases refers to the Taq polymerase enzyme from *Thermus aquaticus* bacteria, extracted from the geothermal aquatic ecosystem of Yellowstone National Park in the United States, during the 1960s. Kary Mullis used this enzyme to develop a molecular biology technique called Polymerase Chain Reaction (PCR) that amplifies ADN fragments at high temperatures on the basis of very small quantities. Kary Mullis was granted a Nobel Prize for PCR and profits from the Taq polymerase enzyme exceed US$200 million a year (Doremus, 1999).

Costa Rica was the first Latin American country to allow bioprospecting in its territory through a non-governmental association with close linkages to the state. The National Biodiversity Institute (INBio) was created in 1991 and signed a formal agreement with the Ministry of Environment and Energy to explore existing biodiversity in protected areas, in exchange for 10 percent from the project budget. It was considered, at the time, an international milestone and economic driver for conservation. At present, INBio does not generate enough income to sustain itself, despite multiple bioprospecting contracts signed with about 30 important companies and universities such as Merck & Co., Syngenta, Strachlyde, among others (Hammond, 2015). Unfortunately, its bilateral bioprospecting model to link national scientific research institutions with foreign companies and research centers (large and medium) failed to produce a viable a long-term sustainable product – after 30 years of research and development. Although the reasons that have led to the downfall of INBio are yet to be evaluated in detail, there is speculation that it could have been due to low return rates and negotiation of relatively low monetary benefits, both explained through economic fundamentals that make it impossible to negotiate fair and equitable contracts (Ruiz, 2015).

Conclusions and final reflections

In Mexico, the fourth most biologically diverse country on the planet, biodiversity is a strategic issue. Approximately 12 percent of the world's species inhabit its terrestrial and marine environments. Biodiversity has an enormous potential to contribute to biotechnology-based research and development, especially in certain industries such as pharmaceuticals, cosmetics, agroindustry and pollution prevention or remediation. Especially as a result of the Nagoya Protocol, Mexico has started to discuss and develop its own national institutional and legal framework on access and benefit sharing from genetic resources, which is to a great extent the same as describing "bioprospecting." This effort requires not only development of administrative and procedural provisions but due consideration to issues such as safe handling of biotechnology, protection of indigenous peoples' rights, technology transfer and, in the case of marine bioprospecting, due consideration to the highly specialized nature of the activity. After defining their set of needs in sectors such as health, food production, environmental remediation and others, Chile, Mexico and Peru should identify opportunities and openings where marine bioprospecting could play an important role in enhancing development. However, it is important to reflect on the lessons learnt from other countries in this field. A chapter or section in the National Plan for Science and Technology: 2018–2024, incorporating marine bioprospecting, could be an important step in broadening the possibilities for Mexico to move further along the road to sustainable development.

Notes

1 In 1967, an enzyme was isolated that allows the joining of DNA fragments from different sources – DNA ligase; in 1970, the first type of enzyme was isolated, capable of cutting DNA at specific sites – nuclease. Finally, in 1977, DNA nucleotide sequencing was developed.
2 The abyssal zone of the ocean is found at depths of 4,000 metres. This region is characterized by its low temperatures, elevated hydrostatic pressure, lack of nutrients and the total absence of light.
3 Metagenomics is defined as the study of microorganisms through their DNA directly from samples that do not require culturing or microbiological isolation. For more information see, Escobar–Zepeda *et al.*, 2015. See also, www.frontiersin.org/articles/10.3389/fgene.2015.00348/full.
4 See, www.un.org/depts/los/convention_agreements/texts/unclos/convemar_es.pdf.
5 See, www.cbd.int/doc/legal/cbd-es.pdf.
6 See, www.conacyt.gob.mx/cibiogem/images/cibiogem/comunicacion/publicaciones/cartagena-protocol-es.pdf.
7 Report of the Intergovernmental Committee for the Cartagena Protocol in Biosafety on the Work of its Third Meeting, Doc. UNEP/CDB/ICCP/3/10, May 27, 2002.
8 Decentralized health authority with technical, administrative and operational autonomy, with the mission of protecting the population against sanitary risks.
9 Decentralized public entity responsible for developing policies and overseeing agricultural sanitary safeguards, protecting aquatic and livestock resources from pests and diseases of quarantine and economic concern, as well as regulating certification of systems to reduce risks of food contamination and their agricultural quality, and facilitate national and international trade of goods of plant and animal origin.

10 It aims to coordinate policies from the Federal Public Administration of Mexico related to biosafety and the production, import, export, mobilization, propagation, consumption and in general, the use and exploitation of Genetically Modified Organisms, their products and sub-products.

11 See, www2.congreso.gob.pe/sicr/cendocbib/con4_uibd.nsf/A7F5BBC269E1220 005257D540063678E/$FILE/RL_28170_ApruebaProtocoloCartagena.pdf.

12 See, www.cbd.int/abs/doc/protocol/nagoya-protocol-es.pdf.

13 The Secretariat of Environment and Natural Resources through the Undersecretary of Development and Environmental Regulation works with indigenous and local communities to strengthen the national implementation of the Nagoya Protocol. Press Release SEMARNAT No. 95/2018 Mexico City, August 28, 2018. See, www.gob.mx/semarnat/prensa/trabaja-semarnat-con-comunidades-indigenas-para-la-implementacion-del-protocolo-de-nagoya?idiom=es.

14 Plankton is a diverse group of free-floating organisms made up of plants, algae, viruses, bacteria and animals. Eighty percent of unicellular organisms that appeared on Earth more than 3 billion years ago were plankton, playing a key role in global climate and biogeochemical cycles.

15 Other similar examples include: TARA Oceans and Tara Oceans Polar Circle (2009–2013), the expedition navigated from the Mediterranean to the Atlantic through the Indian Ocean, Pacific Ocean, Arctic and Antarctic, discovering more than 500,000 new microorganisms – estimating that 95 percent of them continue to be unknown – and 2,600 assembled genomes; and MALASLPINA expedition (2010–2011) where 250 researchers sampled more than 300 stations across the ocean with depths of up to 5,000 metres.

16 The Research Consortium for the Gulf of Mexico was created in 2015 and is formed by approximately 180 researchers from Mexico. Oceanographers, biologists, physicists, chemists and engineers from well-known national research institutions jointly take on the challenge to implement the largest research project in the Gulf of Mexico. The goal is for Mexico to have observation tools, biotechnological development and numerical models to ensure the establishment of contingency plans and mitigation activities in the event of large-scale hydrocarbon spills in the Gulf of Mexico, as well as information to assess environmental impacts.

17 See, www.ggbn.org/.

18 See, http://grbio.org/.

19 See, www.mipatente.com/empresas-spin-off-la-ciencia-tambien-es-negocio/.

20 See, www.pswsa.com/en/.

21 "Bioprospección: Plan Nacional y aproximación al estado actual en Colombia", *Acta Biológica Colombiana*, 2003, Vol. 8, No. 2, p. 73, see, www.invemar.org.co/redcostera1/invemar/docs/3013Plan.pdf.

22 See, https://portal.concytec.gob.pe/index.php/publicaciones/programas-nacionales/item/213-programa-de-valorizacion-de-la-biodiversidad.

23 See, www.smcf.org.mx/avisos/2018/plan-conacyt-ciencia-comprometida-con-la-sociedad.pdf.

References

Blasiak, R., Jouffray, J.B., Wabnitz, C.C.C., Sundström, E., Österblom, H. (2018), Corporate Control and Global Governance of Marine Genetic Resources. *Science Advances* 4(6) eaar5237 DOI: 10.1126/sciadv.aar5237.

Bolívar Zapata, F.G. (Ed.) (2007), *Fundamentos y Casos Exitosos de la Biotecnología Moderna*. 2nd Ed. Mexico, D. F. El Colegio Nacional. With: Academia Mexicana de Ciencias; UNAM, Instituto de Biotecnología, CONACYT, CIBIOGEM.

Costello, M.J., Chaudhary, C., (2017), Marine Biodiversity, Biogeography, Deep-Sea Gradients, and Conservation. *Current Biology*. 27, R511–R527.

Doremus, H. (1999), Nature, Knowledge and Profit: The Yellowstone Bioprospecting Controversy and the Core Purposes of America's National Parks , *Ecology Law Quarterly*. 26(3).

Escobar-Zepeda, A., Godoy-Lozano, E. E., Raggi, L., Segovia, L., Merino, E., Gutierrez-Rios, R. M., Juarez, K., Licea-Navarro, A. F., Pardo-Lopez, L., Sanchez-Flores, A. (2018), Analysis of Sequencing Strategies and Tools for Taxonomic Annotation: Defining Standards for Progressive Metagenomics *Scientific Reports*, 8, 12–34.

Glowka, L., Burhenne-Guilmin, F., Synge, H. *et al.* (1994), *A Guide to the Convention on Biological Diversity*, Environment Policy and Law Paper No. 30, Gland, IUCN, pp. 76–83.

Godoy-Lozano, E., Escobar-Zepeda, A., Raggi, L., Merino, E., Gutierrez Rios, R., Juarez, M., Segovia, K., Licea-Navarro, A., Gracia, F., Sanchez-Flores, A., Pardo-Lopez, L. (2018), Bacterial Diversity and the Geochemical Landscape in the Southwestern Gulf of Mexico. *Frontiers in Microbiology*, 9, 2528.

Gómez-Madrigal, L. S., Moran-Torren, E. F., Méndez-Rivera, J. A. (2014), Bioprospecting and Democracy: A View from the Right to Biological Diversity. *Ciencia Jurídica*. 3(5), 7.

Hammond, E. (2015), Amid Controversy and Irony, Costa Rica's INBio Surrenders Biodiversity Collections and Lands to the State. Third World Network TWN Info Service on Biodiversity and Traditional Knowledge. www.twn.my/title2/biotk/2015/btk150401.htm.

Hurst D., Børresen T., Almesjö L., De Raedemaecker, F., Bergseth, S. (2016), *Marine Biotechnology Strategic Research and Innovation Roadmap: Insights to the Future Direction of European Marine Biotechnology*. Marine Biotechnology ERA-NET.

Melgarejo, L. M., Sanchez, J., Reyes, C., Newmark, F., Santos-Acevedo, M. (2002), *Plan Nacional en Bioprospección Continental y Marina* (propuesta técnica) Bogotá: Cargraphics, (Serie de Documentos Generales INVEMAR No. 11.)

Ruiz, M. (2015), *Genetic Resources as Natural Information. Implications for the Convention on Biological Diversity*, Earthscan from Routledge. London, New York.

Schindel, D. E., Miller, S. E., Trizna, G. N., Graham, E., Crane, A. E. (2016) The Global Registry of Biodiversity Repositories: A Call for Community Curation. *Biodiversity Data Journal* 4: e10293.

Skropeta, D., Wei, L. (2014), Recent Advances in Deep-Sea Natural Products. *Natural Product Reports*. 31, 999.

UNCED and Rio Declaration on Environment and Development. (1992) See, www.unesco.org/education/pdf/RIO_E.PDF.

Wright G., Rochette J., Druel E., Gjerde K. (2016), *The Long and Winding Road Continues: Towards a New Agreement on High Seas Governance*. IDDRI, Paris.

4 Illegal fishing and non-compliance

Rodrigo Oyanedel

Introduction

Conservation and the appropriate management of fisheries largely depends on the level of compliance with regulations (Agnew *et al.*, 2009; Boonstra *et al.*, 2017). Unfortunately, globally, non-compliance with fisheries regulations and norms is a common and extended occurrence (Sumaila, *et al.*, 2006; Agnew *et al.*, 2009). Illegal fishing can be defined as a fishing activity where one or more regulations are breached and/or catches are not reported adequately. It has negative impacts for socio-ecological systems where it ocurrs.

Illegal fishing is linked to the overexploitation of fisheries, destruction of habitats and ecosystems, the use of destructive fishing gear and with the failure of appropriate management systems (Sumaila *et al.*, 2006; Agnew *et al.*, 2009; Raemaekers *et al.*, 2011). Its negative social effects are reflected in tensions and conflicts between users of resources and regulatory authorities (Faasen and Watts, 2007; Lewis, 2015). In addition, the products originating from illegal activities lack sanitary and traceability certifications, which openly threatens public health. Globally, illegal fishing has an estimated economic impact between US$10–23.5 billion, which in terms of landings, represents around 20 percent of marine product catches (Agnew *et al.*, 2009).

Reducing illegal fishing and its impacts involves a complex challenge. There are various intertwined social, economic and ecological factors that should be considered in the search for solutions. This chapter describes different ways in which illegality manifests itself in the fishing sector in Chile, Mexico and Peru. The analysis focuses on the coastal fishing fleets of each country as a starting point to understand and reduce non-compliance and illegal fishing.

The magnitude of illegal fishing and challenges to reduce it in the local context

Mexico

It is estimated that illegal fishing in Mexico represents an additional 45 percent to 90 percent above officially reported landings.[1] This estimation coincides with research by Cisneros-Montemayor and colleagues (2013), which calculated that between 1950 and 2010, the total catches in Mexico were double those officially reported. These numbers reflect illegal catches per se, unreported catches and landings of unregulated species.

The impacts of illegal fishing in Mexico are a constant threat to the sustainability of resources and the sector's competitiveness. Some of the challenges identified to reduce illegal fishing in Mexico include: the *de facto* open access to most fisheries; an extensive and often inaccessible coastline; limited oversight and control capacities by the state of captures and landings; high economic incentives for illegal fishing; limited understanding of the causes of illegal fishing; ineffective control systems and inadequate fines; limited coordination among actors involved in deterring and controlling illegal fishing and low levels of participation of actors in decision-making processes (Cisneros-Montemayor *et al.*, 2013; Cisneros-Montemayor and Cisneros-Mata, 2018).

Chile

In Chile, the National Service for Fisheries and Aquaculture, in charge of overseeing compliance of fisheries regulations, estimated the value of illegal fishing at 300 million US dollars a year (approximately 320 thousand tons).[2] In terms of specific fisheries, one of the most studied cases is the Chilean abalone *Concholepas concholepas* ("*loco*" or "*chanque*") (González *et al.*, 2006; Bandin and Quiñones, 2014; Santis and Chávez, 2015; Adrcu-Cazenave *et al.*, 2017; Oyanedel *et al.*, 2017). For this specie, managed under a special regime of Areas for Management and Exploitation of Benthonic Resources (Castilla and Gelcich, 2017; Gelcich *et al.*, 2010),[3] it is estimated that illegal fishing represents between 70–85 percent of total catches (Oyanedel *et al.*, 2017). The magnitude of illegal and unreported catches for South Pacific hake (*Merluccius gayi gayi*),[4] mixed fishing of common herring (*Strangomera bentincki*) and anchovy (*Engraulis ringen*) has also been estimated (Center of Applied Ecology and Sustainability, 2017).

Challenges to reduce illegal fishing in Chile include: limited state support for supervision and control actions; unreliable traceability to distinguish between legal and illegal catches; high economic dependency on some resources with limited quotas; limited alternative livelihood possibilities for artisanal fishers; poor coordination between state agencies responsible for supervision of fisheries and limited capacity creation platforms to better understand the underlying incentives that promote illegal activities (Bandin and Quiñones, 2014; Davis *et al.*, 2017; Gelcich *et al.*, 2017; Oyanedel *et al.*, 2017).

Peru

Peru is a world leader in fish production, with Peruvian anchovy (*Engraulis ringens*) representing more than 90 percent of landings in the country. A historic analysis of catches for the 1950–2010 period suggests that on average, landings were 24 percent higher than officially reported. This means approximately

82 million tons for that period (Mendo and Wosnitza-Mendo, 2014). The Vice Ministry of Fisheries has estimated that economic losses from illegal activities and informality in general, are between 400 and 500 million US dollars a year.[5]

Informality in the fisheries sector of Peru is one of the major barriers to reducing illegal fishing. It is estimated that there are more than 9,500 vessels without fishing permits or where the information about the technical features of these vessels is unreliable and often contradicts reality.[6] Peru faces additional challenges: limited transparency in activities from both the artisanal and industrial fleets; unreliable control of landings; limited participation of actors in decision-making; ineffective governance structures and systems and capacity to monitor and control fishing.

Recent legislative efforts in Chile and Peru

The threat posed by illegal activities in the fisheries sector has triggered new legislative efforts to provide regulatory and supervisory institutions with new tools in order to reduce noncompliance. Peru and Chile are good examples of how these processes are taking place. Peru has enacted legislation that provides appropriate tools for reduction in illegal fishing. Legislative Decree 1392 regulates the interdiction of illegal fishing activities. It provides authorities with new capacities to disable, confiscate or destroy vessels or any equipment or machinery used for the practice of illegal fishing. Legislative Decree 1392 extends to the operations of processing plants, landings and illegal shipyards. On the other hand, Legislative Decree 1393 seeks to promote the formalization of artisanal fishing activities, through new tools to formalize vessels that do not have a fishing permit. This Decree also applies to vessels whose characteristics differ from the information contained in their registration certificates.[7]

In Chile, the Law for the Modernization and Strengthening of Public Services of the National Fisheries and Aquaculture Service includes a series of provisions to strengthen state control over fishing activities.[8] These include increases in the budget of the National Fisheries and Aquaculture Service, clarification of its competences and functions, development of specific sanctions to be imposed and strengthening inspection processes. Among the critical advances in this law are the definitions of crimes for illegal fishing. This includes targeting post-capture conducts and violations in processing, elaborating, commercializing and storing of products whose origin is questionable or untraceable. The law also includes new requirements applicable to processors and distributors as a condition to register with the National Fisheries and Aquaculture Service. As a result, inspections will be undertaken more effectively along production and commercialization chains and on not only on fishers.

Opportunities and innovations

There are no immediate solutions to address illegal actions in fishing activities. Fundamental changes in the sector and combined efforts among different actors are required. Below are detailed suggested innovations that may assist in designing interventions at different administrative levels.

Innovations in the fisheries management systems

An effective strategy against illegal activities in the fishing sector demands improvements in the management structures of these complex socio-ecological systems. When regulations are perceived as legitimate, fair and reflect the interests of specific actors, the chances for voluntary compliance are improved substantially (Kuperan and Sutinen, 1998; Nielsen, 2003; Keane *et al.*, 2008). It is therefore necessary to innovate towards management systems that support development of regulations which are perceived in a positive manner, thereby increasing voluntary compliance.

Adaptive co-management is an approach to the management of natural resources that emphasizes the need for learning and collaborating to improve the decision-making processes (Armitage *et al.*, 2009). Under these systems, the responsibility for decisions is shared and divided between regulators and those affected by the regulations. On the other hand, in adaptive co-management, actors "learn by doing" which allows conflict resolution based on knowledge and experience generated in situ (Williams, 2011). These systems stimulate participants, engagement in decision-making; access to information is decentralized and is facilitated; finally, the management system responds more appropriately to specific contexts.

Complex situations arise in fisheries management systems when administrative measures are adopted and feedback and responses are received as processes evolve (Berkes, 2002). Polycentric governance systems address these complexities through the application of multiple governance levels, where each unit has certain decision-making independence and autonomy (Ostrom, 2010; Gelcich, 2014). The use of polycentric governance systems may be the vehicle to reduce the power asymmetries of the regulatory system and manage illegal fishing and non-compliance more effectively (Ostrom, 2010).

Concepts such as adaptive co-management and polycentric governance are especially needed to reduce non-compliance with fisheries regulations in countries like Mexico, Peru and Chile. It is common to the analysis of these countries that control and enforcement capacities are insufficient to address the problem of non-compliance. As a result, it is necessary to turn to strategies that increase voluntary compliance with regulations.

Opportunities and research needs

Science has a key role in helping to solve the challenges and obstacles posed by illegal fishing. Estimates of the magnitude of the problem, a better understanding of the causes and solid models to predict the impacts from possible interventions and actions that seek to enhance and improve compliance are needed. Multidisciplinary research needs to be rooted in management needs and specific socio-ecological contexts.

The magnitude of the problem of illegal and/or unreported fishing needs to be accurately estimated. However, the risks for fishers and buyers to acknowledge illegal behaviours, creates a disincentive by fear of reprisals and sanctions (Nuno and St. John, 2015). To overcome this situation various methodologies for indirect surveys have been developed which protect the anonymity of those surveyed. This increases the possibility of sincere responses when sensitive behaviours are investigated. Examples of these methodologies include the Randomized Response Technique (Tracy and Fox, 2012), Unmatched Count Technique (Nuno *et al.*, 2013) and Ballot Box Methods (Bova *et al.*, 2018). Correction methods for catch series also exist, like those presented for Peru, Chile and Mexico (Cisneros-Montemayor *et al.*, 2013; Mendo and Wosnitza-Mendo, 2014; Center of Applied Ecology and Sustainability, 2017). These methods are based on estimates of real catches, through information from different sources, in cases where captures officially reported are recognized to be incomplete. Thereby, estimates are obtained on the real catches in the past. Although there have been efforts to estimate illegal fishing, the scope of these studies needs to be expanded and their accuracy improved to effectively contribute to resource management.

Efforts to estimate illegal fishing must be combined with the creation or improvement of models that explain non-compliance decisions. There are models that explain illegal conducts in economic terms: rational actors decide to fish illegally when the benefits from doing so (profits) outweigh the costs (possibility of detection and magnitude of the fine) (Becker, 1968). This forms the basis of the deterrence model that seek to reduce illegal conducts by increasing the costs and/or reducing the benefits from illegal fishing. However, various problems emerge when applying this model to fisheries in Chile, Peru and Mexico. First, resources for inspections are insufficient and state agencies lack the tools, human capacity and technologies needed to effectively detect non-compliance (Davis *et al.*, 2017; Oyanedel *et al.*, 2017). Second, the difficulty of arresting fishers in situ and low fines mean that the costs of operating illegally are not affected and there is no deterrent to modify these conducts (Kuperan and Sutinen, 1998).

Given that the deterrent approach to reduce illegal fishing seems to be ineffective, new management and compliance models need to be developed. Normative factors such as the legitimacy of regulations (Nielsen, 2003; Keane *et al.*, 2008), or social factors such as the perception that certain conducts are right given that the majority does it, are decisive with regards to individual decisions to comply or not with regulations (Hatcher *et al.*, 2000; Cinner, 2018). Interdisciplinary science can play a key role for unravelling different factors that underscore the response of fishers to regulations in the specific contexts of these three countries. The construction of new models for a better understanding of how to intervene in order to modify incentives for users, is key to reduce illegal fishing (Keane *et al.*, 2008).

Technological innovations

A common lesson which emerges in Chile, Mexico and Peru is that there is a need to increase the use of technological tools in the fisheries sector. Technology can assist, for example, in the implementation of integrated traceability systems. The objective of traceability is to increase transparency in the supply chain, differentiate legal from illegal products and, in certain cases, increase value by means of competitive differentiation (Moretti *et al.*, 2003). This technology can assist enforcement processes, through the identification of potential "bottlenecks" in the supply chain, where products from different ports and fishers converge. Focusing on these points can be more efficient in economic terms and regarding the effects of enforcement actions.

One of the most recent technological developments in the effort to address illegality in fishing is the Global Fishing Watch platform (Kroodsma *et al.*, 2018). This initiative uses cutting-edge technology to visualize and follow the activities of fishing vessels around the world, promoting sustainable use of marine resources through transparency. Peru has been a leader in adopting this technology by sharing the Vessel Monitoring System data from all its fleet to GFW. This is a promising tool to reduce illegal fishing, but it has not been applied yet to coastal fleets that lack tracking devices. It will be necessary to understand the incentives, barriers and enabling conditions of these coastal fleets to adopt tracking devices and scale-up the use of technologies such as GFW.

Conclusions, reflections and perspectives

To understand and reduce illegal fishing in a local context is a process that requires multi-dimensional efforts, varied disciplines and different level interventions. The need to improve or reform their fisheries management systems is common for these countries. Adaptive co-management and polycentric systems with structures allowing for the participation of actors are bridges towards greater levels of voluntary compliance with regulations. More effective management systems also contribute to reducing informality in the fisheries sector and increase transparency. It is only through a constant and adaptive learning effort that the necessary conditions to improve compliance will be understood.

On the other hand, more active and coordinated enforcement and control systems are necessary. In Chile, Peru and Mexico, the resources for supervisions are not adequate: there needs to be more staff and use of technologies, knowledge and capacities. In addition, there is a low coverage (mainly for artisanal fleets) of traceability systems that monitor fisheries and products along the supply chain. This implies often limited or non-existent information as to the origin, safety and legality of the products in the market. Overlapping and often competing competences among government agencies is also common in most of Latin America. From authorities that regulate extraction to agencies in charge of food safety, fishing activities are not nestled in a central regulatory or administrative body. This diversity of agencies inevitably causes a coordination problem for enforcing and controlling compliance. It is fundamental to implement effective mechanisms and coordination channels between the agencies that control fishing activities at all levels.

It is through the combination of increased voluntary compliance and better enforcement that illegal fishing activities can be reduced. These two approaches must be appropriately coordinated, adapting to local contexts and including key actors in the design, implementation and monitoring of different interventions. Efforts to combat illegal fishing only based on fines and sanctions have the unintended consequence of excluding actors from the system, destroying trust and reducing the legitimacy of the management system. The often politically motivated narrative of "combating illegal fishing" runs the risk of separating regulators and the regulated into two trenches. On the other hand, it is naïve to think that voluntary compliance is enough to reduce illegal fishing and move towards sustainable development. The challenge in fisheries management for these countries lies in how to combine and coordinate these two approaches.

The cases described in this chapter are not a comprehensive representation of all forms of illegal fishing and non-compliance in Mexico, Peru and Chile. Different illegal practices have been left aside, such as fishing in marine protected areas, or fishing with prohibited gear. However, based on the experiences presented and opportunities and innovations outlined, new ways to address the problem of non-compliance emerge. A new approach to reduce illegal fishing is fundamental: one based on coordinated and focused enforcement, adaptive processes and new governance structures that include all relevant actors, whether they are legal or not.

Notes

1 Environmental Defense Fund (2013), *La Pesca Ilegal e Irregular En Mexico: Una Barrera a La Competitividad*. Available at, https://mexico.edf.org/sites/mexico.edf.org/files/pescailegalfinal-07–06–17.pdf.
2 See, www.economiaynegocios.cl/noticias/noticias.asp?id=455033.
3 These areas are an access regime which assigns exploitation rights to a group or association of artisanal fishers in a specific area.
4 CEDEPESCA. *Estudio de Estimación del Sub Reporte Artesanal en la Pesquería de Merluza Común.*
5 See,http://eltiempo.pe/atkins-en-peru-400-millones-de-dolares-se-pierden-por-la-pesca-ilegal/.
6 PRODUCE. Anuario Estadistico Pesca y Acuicultura 2012. Available at, www.produce.gob.pe/documentos/estadisticas/anuarios/anuario-estadistico-pesca-2012.pdf.
7 Sociedad Peruana de Derecho Ambiental (2018), *Análisis Legal sobre los Decretos para Combatir la Pesca Ilegal para iniciar un Nuevo Proceso de Formalización*. Available at, www.actualidadambiental.pe/?p=51886.
8 Bulletin10483–21.

References

Agnew, D. J. *et al.* Estimating the Worldwide Extent of Illegal Fishing. *PLoS One* 4 (2), (2009).

Andreu-Cazenave, M., Subida, M. D., Fernandez, M. Exploitation Rates of Two Benthic Resources Across Management Regimes in Central Chile: Evidence of Illegal Fishing in Artisanal Fisheries Operating in Open Access Areas. *PLoS One* 12, e0180012 (2017).

Armitage, D. R. *et al.* Adaptive Co-Management for Social-Ecological Complexity. *Front. Ecol. Environ.* 7, 95–102 (2009).

Bandin, R., Quiñones, R. A. Impacto de la Captura Ilegal en Pesquerías Artesanales Bentónicas bajo el Régimen de Co-manejo: el Caso de Isla Mocha, Chile. *Lat. Am. J. Aquat. Res.* 42, 547–579 (2014).

Becker, G. S. Crime and Punishment: An Economic Approach. *J. Polit. Econ.* 76, 169–217 (1968).

Berkes, F. *Navigating Social–Ecological Systems.* (Cambridge University Press, 2002). doi:10.1017/CBO9780511541957.

Boonstra, W. J., Birnbaum, S., Björkvik, E. The Quality of Compliance: Investigating Fishers' Responses Towards Regulation and Authorities. *Fish Fish.* 18, 682–697 (2017).

Bova, C. S., Aswani, S., Farthing, M. W., Potts, W. M. Limitations of the Random Response Technique and a Call to Implement the Ballot Box Method for Estimating Recreational Angler Compliance using Surveys. *Fish. Res.* 208, 34–41 (2018).

Castilla, J. C., Gelcich, S. Management of the Loco (*Concholepas concholepas*) as a Driver for Self-Governance of Small-Scale Benthic Fisheries in Chile. 1989, 441–452 (2007).

CEDEPESCA. *Estudio de Estimación del Sub Reporte Artesanal en la Pesquería de Merluza Común* (2016) Available at, http://cedepesca.net/wp-content/uploads/2016/11/2016-09_CeDePesca_Estimacion-de-la-cota-minima-de-subreporte-artesanal-merluza-comun-Chile-2.pdf.

Center of Applied Ecology and Sustainability (CAPES). Metodología para la Estimación de las Capturas Totales Anuales Históricas. Caso de Estudio: Pesquería de Sardina Común y Anchoveta V–X Región. 1–4 (2017).

Cisneros-Montemayor, A. M., Cisneros-Mata, M. A. A. Medio Siglo de Manejo Pesquero en el Noroeste de México, el Futuro de la Pesca como Sistema Socioecológico. *Relac. Estud. Hist. y Soc.* 39, 99 (2018).

Cisneros-Montemayor, A. M., Cisneros-Mata, M. A., Harper, S., Pauly, D. Extent and Implications of IUU Catch in Mexico's Marine Fisheries. *Mar. Policy* 39, 283–288 (2013).

Cinner, J. E. How Behavioral Science can Help Conservation. *Science* (80) 362, 889–891 (2018).

Davis, K. J. *et al.* Why are Fishers not Enforcing Their Marine User Rights? *Environ. Resour. Econ.* 67, 661–681 (2017).

Environmental Defense Fund. *La Pesca Ilegal e Irregular En Mexico: Una Barrera a La Competitividad.* (2013).

Faasen, H., Watts, S. Local Community Reaction to the 'No-Take' Policy on Fishing in the Tsitsikamma National Park, South Africa. *Ecol. Econ.* 64, 36–46 (2007).

Gelcich, S. *et al.* Fishers' Perceptions on the Chilean Coastal TURF System after Two Decades: Problems, Benefits, and Emerging Needs. *Bull. Mar. Sci.* 93, 53–67 (2017).

Gelcich, S. Towards Polycentric Governance of Small-Scale Fisheries: Insights from the New 'Management Plans' Policy in Chile. *Aquat. Conserv. Mar. Freshw. Ecosyst.* 581, 575–581 (2014).

Gelcich, S. *et al.* Navigating Transformations in Governance of Chilean Marine Coastal Resources. *Proc. Natl. Acad. Sci.* 107, 12794–16779 (2010).

González, J. *et al.* The Chilean Turf System: How is it Performing in the Case of the Loco Fishery? *Bull. Mar. Sci.* 78, 499–527 (2006).

Hatcher, A. *et al.* Normative and Social Influences Affecting Compliance with Fishery Regulations Normative. *Land Econ.* 76, 448–461 (2000).

Keane, A. M., Jones, J. P. G., Edwards-Jones, G., Milner-Gulland, E. J. The Sleeping Policeman: Understanding Issues of Enforcement and Compliance in Conservation. *Anim. Conserv.* 11, 75–82 (2008).

Kroodsma, D. A. *et al.* Tracking the Global Footprint of Fisheries. *Science* 359, 904–908 (2018).

Kuperan, K., Sutinen, J. G. Blue Water Crime: Deterrence, Legitimacy, and Compliance in Fisheries. *Law Soc. Rev.* 32, 309 (1998).

Lewis, S. G. Bags and Tags: Randomized Response Technique Indicates Reductions in Illegal Recreational Fishing of Red Abalone (*Haliotis rufescens*) in Northern California. *Biological Conservation* 189, 72–77 (2015).

Mendo, J., Wosnitza-Mendo, C. Reconstruction of Total Marine Fisheries Catches for Peru: 1950–2010. *Fish. Centre. Univ. Br. Columbia. Work. Pap. #2014–21* 23 (2014).

Moretti, V. M., Turchini, G. M., Bellagamba, F., Caprino, F. Traceability Issues in Fishery and Aquaculture Products. *Veterinary Research Communications* 27, 497–506 (2003).

Nielsen, J. R. An Analytical Framework for Studying: Compliance and Legitimacy in Fisheries Management. *Mar. Policy* 27, 425–432 (2003).

Nuno, A., St John, F. A. V. How to Ask Sensitive Questions in Conservation: A Review of Specialized Questioning Techniques. *Biological Conservation* 189, 5–15 (2015).

Nuno, A., Bunnefeld, N., Naiman, L. C., Milner-Gulland, E. J. A Novel Approach to Assessing the Prevalence and Drivers of Illegal Bushmeat Hunting in the Serengeti. *Conserv. Biol.* 27, 1355–1365 (2013).

Ostrom, E. Polycentric Systems for Coping with Collective Action and Global Environmental Change. *Glob. Environ. Chang.* 20, 550–557 (2010).

Oyanedel, R., Keim, A., Castilla, J. C., Gelcich, S. Illegal Fishing and Territorial User Rights in Chile. *Conserv. Biol.* 32, 619–627 (2017).

PRODUCE. Anuario Estadistico Pesca y Acuicultura 2012. Available at, www.produce.gob.pe/documentos/estadisticas/anuarios/anuario-estadistico-pesca-2012.pdf.

Raemaekers, S. *et al.* Ocean & Coastal Management Review of the Causes of the Rise of the Illegal South African Abalone Fishery and Consequent Closure of the Rights-Based Fishery. *Ocean Coast. Manag.* 54, 433–445 (2011).

Santis, O., Chávez, C. Quota compliance in TURFs: An Experimental Analysis on Complementarities of Formal and Informal Enforcement with Changes in Abundance. *Ecol. Econ.* 120, 440–450 (2015).

Sociedad Peruana de Derecho Ambiental (2018), Análisis Legal sobre los Decretos para Combatir la Pesca Ilegal e Iniciar un Nuevo Proceso de Formalización. Available at, www.actualidadambiental.pe/?p=51886.

Sumaila, U. R., Alder, J., Keith, H. Global Scope and Economics of Illegal Fishing. *Mar. Policy* 30, 696–703 (2006).

Tracy, P. E., Fox, J. A. The Validity of Randomized Response For Sensitive Measurments. *Am. Sociol. Rev.* 46, 187–200 (2012).

Williams, B. K. Adaptive Management of Natural Resources: Framework and Issues. *J. Environ. Manage.* 92, 1346–1353 (2011).

5 Extractive industries in coastal and marine zones

Andrea Cuba

Introduction

Coastal and marine zones have always been essential for human wellbeing and in some cases, survival, particularly in countries like Peru, where 60 percent of its population and growing, lives near or on the coastline (INEI, 2017). Coastal and marine zones are geomorphologic spaces with important economic, social and cultural dimensions and which provide a wide range of services to society, both directly and indirectly. They are also a source of ecosystem services, food security, and strongly influence the national economy and international trade through activities therein (UNEP-WCMC, 2011).

These features also make coastal and marine zones extremely vulnerable to the impact of economic activities, whether undertaken within their geographical limits, close by or elsewhere. They are often exposed to irrational and excessive uses by human beings and industries in particular. To date, 29 percent of fish stocks worldwide are over-exploited and 61 percent are fully exploited (OCEANA, 2016). In general, the increased pressure caused by fishing of all sorts has produced a decline in marine wildlife (Hooker Mantilla, 2011). Overfishing, activities which do not comply with environmental and sustainability laws and regulations, marine pollution from multiple sources and openly illegal extractive activities, all have a very visible impact on these coastal and marine spaces.

The reality of extractive industries in coastal and marine zones in Peru

The distinct oceanographic features of the Peruvian Sea include a complex system of currents that leads to one of the most important upwelling systems in the world.[1] At the same time, Peru has an extractive industry that has been growing from year to year in the area of mining, forestry, fisheries and hydrocarbons. In 2017, mining alone represented around 10 percent of the Gross Domestic Product (GDP) and nearly 62 percent of the total value of Peruvian exports (MINEM, 2017), while during the same year fisheries increased by 82 percent (Sociedad Nacional de Pesquería, 2017). It is quite a paradox that given the immense economic value that these industries represent, out of the 127 socio-environmental conflicts reported in Peru until August 2018, 82 were directly related to mining activities and 17 to hydrocarbon activities.[2] Conflicts regarding fisheries are much less visible and take place marginally, usually between artisanal fishers and industrial fleets or when hydrocarbon-related exploitation is awarded within 5 miles of the coastal border.[3] Conflicts have also occurred due to limited and delayed certification of artisanal piers and the creation of marine protected areas, among others.[4]

On the other hand, Coastal and Marine Spatial Planning is a tool which is only recently being considered and developed in the country.[5] In 2015, the Peruvian State took an important step with the publication of Guidelines for Integral Management of Marine and Coastal Zones,[6] which promotes the sustainable use of natural resources and conservation of coastal-marine ecosystems.[7] These Guidelines have only recently been implemented in some regions of the Peruvian coast, such as Piura, in the northern part of the country.[8] Another useful management tool that Peru has postponed is Marine Protected Areas. Until 2010, there was only one Marine (and coastal) Protected Area – the Paracas National Reserve, created in 1975. With the creation of the San Fernando National Reserve, and the National Reserve System of Islands, Small Isles and "Guano" Points, the number has increased to three Marine Protected Areas.[9]

The challenges of fishing in Peru

Fishing in Peru represents an important source of food for many people and is a big employer. Almost 83,000 jobs are directly linked to fishing whilst 25,000 have an indirect link (Sociedad Nacional de Pesquería, 2017a). Its contribution to GDP is on average between 1.5 percent to 2 percent, though this varies from year to year, and it also contributes to 7 percent of Peru's exports (Sociedad Nacional de Pesquería, 2017b).

The state determines fisheries management systems based on the type of fisheries and the conservation situation of targeted resources. The management instruments are regulated in the General Fisheries Law,[10] its Regulation[11] and Fisheries Management Regulations. However only ten species, including tuna, eel, mackerel and anchovy, have a specific Fishery Management Regulation applicable to them.

The main criteria used by the Ministry of Production when granting access to a fishery, is based on exploitation status, which is determined by technical information provided by the Institute for the Sea of Peru, which is part of the Ministry of Production. However, given the existing informality, especially within small-scale fishing, precise and reliable data and information is hard to obtain and validate. Additionally, Peruvian legislation does consider the criteria of *over*exploitation as a defining condition, which precludes this problem from being formally addressed as part of the current legal framework and remedies.

In the specific case of small-scale fishing the existing open access regime for most marine resources puts excessive pressure on some of these and causes unsustainable practices to prevail with no legal recourse available to address this situation. Informality furthermore hampers any inspection and effective sanctions for illegal activities. According to the First National Census of Artisanal Fisheries in the Marine Area, only 52 percent of fishers have some type of certification and between 70 percent and 80 percent have an authorization for individual fishing activities (INEI, 2012).

To address this situation, in 2016 the state launched the Artisanal Fishery Formalization Program. This effort has had a slow start due to climate events and destruction caused by El Niño in 2017 and problems regarding outreach and information about this process reaching fishers.

With regards to industrial fisheries, anchovy (*Engraulis ringens*) is by far the most important resource. It represents approximately 86 percent of landings for indirect human consumption (MINAM, 2015). Though fortunately resilient over time, anchovy is constantly on the verge of depletion or overexploitation. Although the anchovy fishery was re-structured in 2014 with the establishment of new fishing zones,[12] illegal fishing and unabated overexploitation continue to put pressure on this resource. In the case of the industrial fishery of anchovy "there continues to be deficiencies in monitoring and control, with inefficient regulations to prevent and mitigate juvenile catches, effects on spawning and incidental catches of other species" (MINAM, 2012).

In 2008, Peru took an important step to address these problems and challenges by including the crime of "illegal extraction of aquatic species" in the Penal Code.[13] This helps in the effort to provide greater legal protection for marine resources by creating a disincentive for catches below minimum sizes and banning prohibited fishing methods and equipment (e.g. explosives, chemical products or unregulated nets).[14] The Penal Code also sanctions with effective prison time the extraction of species from Natural Protected Areas – without a valid authorization.[15] In addition to this, the creation of Specialized Environmental Prosecutor Offices in the different regions of Peru, has also contributed to streamlining and more effective legal responses to crimes committed against hydrobiological resources and coastal areas in general.

The case of mining in Peru and its impacts on the coast and seas

Seemingly unrelated to fishing, mining is arguably the most important and profitable extractive industry in Peru. The Ministry of Energy and Mines has granted concessions rights over a fifth of the Peruvian territory to mining activities (Grupo Propuesta Ciudadana, 2014). Many of these rights cover areas near the coast and some cover the Peruvian sea.[16] One of the main impacts from this industry comes from wastes and discharges into bodies of water – rivers, streams, lakes or directly the sea. These discharges contain toxic metals such as lead, cadmium and mercury. If these metal residues are not processed and controlled, they can become dangerous both for the marine ecosystem and human health. Filtrations and leachate from abandoned mine tailings also cause deterioration of water resources. Marine areas in Peru that present the greatest concentration of metal traces in their sediments are the Bay of Ferrol in Chimbote and the Bays of Callao and Pisco (MINAM, 2015).

Two sources of pollution are present in coastal and marine areas. The first comes from formal and authorized mining activities but which do not comply with existing environmental management tools such as environmental impact assessments; the second comes from openly illegal and informal mining which does not follow any existing legal environmental protection parameter.[17]

With regards to the former, during the second trimester of 2018, the Environmental Assessment and Control Agency determined that 69 percent of total violations to environmental standards related to non-compliance of environmental management tools by mining operators (OEFA, 2018). Though sanctions have been imposed and deterrent actions taken, the fact is environmental damage often occurs and remediation and ex post clean ups are mostly ineffective.

When discussing informal mining in Peru, this inevitably leads to small-scale and artisanal mining (SINIA, 2017). In 2012, the state enacted a series of regulations to incentivize formalization and registration by small-scale and artisanal miners, through simplified administrative procedures and benefits for registration compliance.[18] The Environmental Management Instrument for the Formalization of Small-Scale and Artisanal Mining Activities,[19] aims to engage and allow collaboration with informal miners and create appropriate mechanisms and support for complying with environmental standards, suspending temporarily penal action and sanctions as a means to develop an enabling environment for formalization.

However, illegal miners still remain very much active as do those informal miners who have no real intention of becoming formalized and subject to existing environmental legal and regulatory frameworks. This group of miners causes most of the damage to coastal and marine areas (MINAM, 2016). According to official information available from the Ministry of Energy and Mining, environmental liabilities (e.g. tailing ponds, polluted ecosystems) left from mining activities in Peru reached 5,556 situations by 2010.[20] Although legislation promotes the recovery of degraded areas through the development and approval of a Plan for the Recovery of Environmental Impacts Generated by Illegal Mining, to date this the specific instrument has not been approved.[21]

In the convergence between mining and fisheries and coastal and marine zones, Specialized Environmental Prosecutor Offices have become key state institutions to combat illegal mining activities. In 2012, illegal mining crimes were also incorporated into the Penal Code,[22] with sanctions of up to ten years in prison.[23] Likewise, in 2012 the Public Ministry was authorized to undertake interdiction action in coordination with Specialized Environmental Prosecutor Offices.[24]

Challenges for the exploration and exploitation of hydrocarbons in Peru

The north-western coast of Peru is the oldest and most exploited hydrocarbon region in the country (IMARPE, 2010). At present, there are at least 13 oil lots located in the Peruvian sea, under a formal concession (right) granted to national and international companies. Over the years, evidence has accumulated regarding the impact of seismic prospecting on marine life, which causes waves that affect the auditive capacity of whales, dolphins and schools of fish, in some cases even resulting in their death (Sierra Praeli, 2018). Most certainly, during the exploitation stage of hydrocarbons, there is a high risk of pollution of the marine ecosystem due to spills and related accidents. It has been shown time and again that this often causes irreversible damage to marine biodiversity, coastal ecosystems and communities living in the areas of influence.

According to the Law for Hydrocarbons,[25] the areas licensed by PERU-PETRO S.A.[26] to companies are granted based on the "hydrocarbon potential, geographic zone, guaranteed minimum work program and area where exploration and exploitation of hydrocarbon activities will effectively be undertaken" (PERUPETRO, 2019). Environmental considerations or risks are not included within these criteria. Likewise, this law does not distinguish between licenses for exploration and exploitation in the sea or on land, which creates an additional problem regarding safeguards required for specific and different marine or land ecosystems.

History repeats itself: over time companies have been granted concessions in coastal and marine areas with limited, if any, conditions regarding the specific protection of these ecosystems. In 2018, for example, five regulations were enacted which authorized the exploration and exploitation of five oil lots in the seas of Tumbes, Piura, Lambayeque and Ancash.[27] These regulations were highly controversial for various factors, including limited citizen participation in the decision to grant these lots and consideration to environmental risks derived from oil well infrastructure constructed at sea. As a result of pressure from the media and the public, a few months later the Ministry of Energy and Mines repealed these regulations. However, this case is important as it shows how oil lots are frequently granted without a prior comprehensive analysis of the risks involved, particularly when they may affect vulnerable ecosystems and sustainability of converging activities such as fisheries.

Oil spills without rapid or with inadequate responses, as for example, when dispersants are used to eliminate only ocean surface pollution, but not sea on the floor, can have devastating effects on marine habitats. In these cases, the Directorate General of Captaincies and Coast Guard of the Peruvian

Navy has a National Contingency Plan to control and mitigate oil spills and accidental discharges of other harmful substances into the seas, rivers and lakes.[28] Likewise, each operator (i.e. company undertaking an exploration of exploitation activity) must have its own approved Contingency Plan. In reality, plans are inadequate or never up-dated, emergency drills are non-existent and, even in the hydrocarbon industry, informality also affects the possibility of having appropriate and effective responses.

Protection measures in coastal and marine zones of other countries

Chile

Chile has a great diversity of land, marine and coastal ecosystems, including glaciers, rivers, lakes, islands and wetlands, which hold nearly 30,000 plant and animal species, fungi and bacteria (Ministerio de Medio Ambiente de Chile, 2014). This rich biodiversity has been over time subjected to human-induced pressures and overexploitation.

Fisheries in Chile contribute 0.4 percent to the national GDP (FAO, 2012). In comparison, mining represents nearly 9 percent of the national GDP (Banco Central de Chile, 2017). Additionally, the aquaculture industry has been growing, with trout and salmon being the main exports.

In parallel to economic development, these industries have taken their toll on the environment – on ecosystems and species in particular. Out of 33 major fisheries in 2012 in Chile, 15 were being fully exploited, 10 overexploited, 3 had collapsed and the status of 5 was uncertain due to limited information and data (Subsecretaría de Pesca y Acuacultura del Gobierno de Chile, 2012). In addition to these pressures, the presence of both heavy metals and sediments was noted, mainly caused by mining extractive activities and its discharges to the ocean and in coastal ecosystems (Ministerio del Medio Ambiente de Chile, 2014). As a result of this situation, and with the aim of protecting the marine and coastal zone, the Chilean government has created and implemented a series of instruments, including marine concessions, management tools and marine protected areas (Maguiña, 2009).

Marine concessions are granted and confer rights to legal or natural persons who seek to develop and undertake projects and enterprises which involve beaches, the ocean floor or portions of marine waters which are otherwise considered national goods for public use.[29] To obtain a concession, the project (e.g. extraction of minerals or development of industrial complexes) must pass through the Environmental Impact Assessment System.

Chile has been recognized for introducing a fisheries management system based on access rights for the sustainable extraction of its benthonic resources. This system was originally set up as a voluntary mechanism for artisan fishers, and in the early 1990s it was formally recognized by the state. This management system helps to balance the pressures of fishing in an open access regime, by imposing restrictions on access and how much can be extracted in a specific fishery.

Marine protected areas have also proved to be an effective tool to address inappropriate use of marine resources, and also to organize and mitigate effects of extractive and non-extractive activities in the sea. By 2018, Chile had 40 percent of its exclusive marine economic zone under some form of environmental protection (FAO, 2018). Marine protected areas in Chile are divided into four categories: marine reserves which protect small repro-duction, spawning zones and important zoological interests; natural sanctuar-ies which enable the study and research of natural formations which are of particular interest for the state; coastal and marine zones destined for multiple uses and sustainable exploitation; and marine parks which are strictly protected and closed to any extractive activity, except consumption by local and coastal indigenous communities (Ministerio del Medio Ambiente, 2014).

Mexico

Mexico is part of a megadiverse group of countries, hosting approximately 70 percent of the world's diversity of species (ONU MEXICO, 2016).

Related to hydrocarbon exploitation, the petrochemical industry has led the way for economic development in Mexico. Although economic rents from hydrocarbons have been in steady decline over the years, its impacts have been lasting. For example, the spillage of more than 900 million liters of crude oil in 2010, when the oil platform Deepwater Horizon exploded in the Gulf of Mexico (in US territorial waters), became the most catastrophic disaster in the history of the environment, comparable to Chernobyl and the Exxon Valdez tragedies, with effects that are still felt in the Gulf Coast states of the US and the eastern coastal border of Mexico and beyond (Cousteau, 2015).

As of 2014, Mexico already had 174 federal protected areas, including a wide range of ecosystems, with over 25,384,818 hectares of coverage, repre-senting 12.85 percent of the national territory. Of this percentage, 2.42 percent represents marine zones, with a total of 34 marine and coastal protected areas.[30] In 2017, Mexico created the largest marine protected area in North America (Revillagiedo Archipelago in the Pacific), with a coverage of 14.8 million hectares,[31] which seeks to promote the conservation of marine ecosystems and their biodiversity, through direct community parti-cipation in management and sustainable use. Both Mexico and Chile have advanced considerably in the establishment of payment for ecosystem services schemes, to create opportunities for local populations to benefit from the restrictions established by marine protected areas.

Unlike Peru, Mexico has developed marine ecological planning, which considers the ecologic and social attributes of a marine space in order to organize the coexistence of productive activities, provide effective environ-mental protection and facilitate conflict resolution between different users through continuous dialogue (Espinoza-Tenorio, et al., 2014).

Conclusions and vision for the future

This chapter has sought to transmit the idea that both conservation of marine and coastal zones and the prevention of social conflicts demand careful balancing between extractive industry interests and those of local, coastal communities and the state in general. Although Peru has made progress with regards to legal regulations and guidelines for coastal and marine spatial planning, implementation of effective measures and effective actions on the ground are still limited. Convergence of uses and interests has resulted in conflicts and often erosion of coastal marine borders, continuous pollution and depletion of marine resources and fisheries.

Likewise, the inadequate coordination between different competent authorities and extended informality in the extractive industries in Peru, further increases the complexity of the problems and affects the possibility of exercising regulatory and control functions. Furthermore, the inadequate protection of marine ecosystems is also reflected in limited use of effective conservation and management tools such as marine protected areas.

During the last few years, the Peruvian state has been discussing the creation of the Tropical Pacific Ocean Reserve which would include five massive marine biodiversity zones: four located within five nautical miles of the Peruvian Sea, and a fifth, the Mancora Bank, 40 miles from the coast (Sierra Praeli, 2018). However, this reserve has not yet been created, due to unsolved conflicts, mainly with regards to converging rights of fishers overlapping with hydrocarbon lots allocated in the Peruvian sea (García, 2018). This conservation tool must be seen as a means to better manage common goods and ensure converging interests can coexist in a balanced manner.

Chile and Mexico showcase conservation instruments which can effectively be deployed to achieve conservation of marine and coastal zones objectives. They have been useful to mitigate the impacts of different extractive activities according to specific characteristics of the targeted area and the type of activity planned.

In order to reach a peaceful coexistence between the extractive industry and protection of coastal and marine zones in Peru – and in any other parts of the world – a first step is to develop a strong legal framework which includes, for example, marine spatial planning, environmental impact assessments and marine protected areas, among other tools which may be necessary under particular circumstances. A second step requires clearly defined competences among authorities to prevent overlaps and uncertainty with regards to "who-is-in-charge-of-what". Third, and this may be more relevant to certain countries, addressing and eliminating informal activities is critical to ensure appropriate management and conservation interventions. This does not imply necessarily eliminating certain extractive activities in coastal and marine areas altogether – this is an option for countries such as Costa Rica – but to regulate them in such a way that they have a positive or neutral impact conservation-wise.

Notes

1 See, Humboldt Current Large Marine Ecosystem Project, report available at, http://humboldt.iwlearn.org/es/informacion-y-publicacion/AbstractProductivity.pdf.
2 Report on Social Conflicts by the Ombudsman's Office (2018).
3 The case of artisanal fishers in Northern Peru against the oil industry is emblematic. See, García, R. *Pescadores Protestan contra Concesiones Petroleras en el Norte del Perú*. Available at, https://es.mongabay.com/2018/04/pescadores-paro-petroleo-oceanos-peru/.
4 See, www.actualidadambiental.pe/?p=49411.
5 The SPINCAM Project (South Pacific Data and Information Network in support of Integrated Coastal Area Management) was designed under the coordination of IOC-UNESCO and the Permanent Commission for the South Pacific, PCSP, to establish an indicator framework for integrated coastal area management (ICAM) at the national and regional levels for countries in the South Pacific.
6 Approved through Ministerial Resolution 189–2015-MINAM, August 5, 2015.
7 Strategic Guideline 4.
8 In 2014, the Regional Government approved an "Integrated Management Plan for Resources in the Coastal-Marine Zone of Sechura".
9 See, www.actualidadambiental.pe/pacificotropical/.
10 Decree law 25977, General Law of Fisheries, December 22, 1992.
11 Supreme Decree 012–2001-PE, Regulation of the General Law of Fisheries, March 14, 2001.
12 The publication of Supreme Decree 005–2012-PRODUCE, Regulation on Fisheries Management for Anchovy and White Anchovy Resources, establishes an exceptional regime of reserved zones for direct human consumption.
13 Approved through Legislative Decree 635, Penal Code, April 8, 1991.
14 Article 308B of the Peruvian Penal Code, approved through Legislative Decree 635, establishes between three and five years of prison time for these violations.
15 Article 309 of the Peruvian Penal Code, approved through Legislative Decree 635, establishes between four and seven years of prison time for this type of violation.
16 Available at: www.ingemmet.gob.pe/web/guest (accessed January 2019).
17 Illegal mining refers to mining activities in forbidden areas in the absence of an authorization granted by the authority, while informal mining is also conducted without an authorization, but currently subject to a formalization process.
18 Legislative Decree 1105, that establishes provisions for the Small-Scale and Artisanal Mining Formalization Process, April 19, 2012.
19 According to Legislative Decree 1336, that establishes provisions for the Integral Mining Formalization Process, January 6, 2017.
20 Ministerial Resolution 371–2010-MEM-DM, updates the Initial Inventory of Mining Environmental Liabilities approved through R.M. 290–2006-MEM/DM, August 28, 2010.
21 Legislative Decree that regulates the interdiction of illegal mining throughout the Republic and establishes complementary measures, February 18, 2012.
22 Through Legislative Decree 1102, that incorporates illegal mining offences in the Penal Code, February 29, 2012.
23 Article 307A of the Peruvian Penal Code, approved through Legislative Decree 635.
24 By Legislative Decree 1100, that regulates the interdiction of illegal mining throughout the Republic and establishes Complementary Measures, February 18, 2010.

25 Law 26221, Organic Law that governs hydrocarbon activities in the national territory, August 20, 1993.
26 Private enterprise of the Peruvian State, that grants property rights on hydrocarbons extracted for the purpose of exploration and exploitation contracts and simple exploitation contracts.
27 Supreme Decrees 006–2018-EM, 007–2018-EM, 008–2018-EM, 009–2018-EM and 010–2018-EM, that correspond to lots Z-64, Z-65, Z-66, Z-67 y Z-68 respectively, March 24, 2018.
28 Supreme Decree No. 051-DE/MGP, national contingency plan to control and mitigate hydrocarbon spills and other harmful substances into the seas, rivers and navigable lakes, August 2, 1993.
29 Regulation for Maritime Concessions of Chile, Decree (M) No. 660 of 1988.
30 See,SEMARNAT(2013):www.gob.mx/semarnat/articulos/mexico-primer-lugar-en-la-proteccion-de-areas-marinas.
31 Available at: www.oceansentry.org/ (accessed in January, 2019).

References

Banco Central de Chile (2017). *Informe de Cuentas Nacionales de Chile, Cuarto Trimestre de 2017.*

Cousteau, J.M. *Ocean Futures Society.* (2015), Available at, www.oceanfutures.org/news/blog/Derrame-de-petroleo-del-Deepwater-Horizon-5-anos-de-secuelas (Accessed January 10, 2018).

Espinoza-Tenorio, A., *et al.* (2014), El Ordenamiento Ecológico Marino en México: un Reto y una Invitación al Quehacer Científico. *Lat. Am. J. Aquat. Res.,* 42(3): 386–400, 2014El q.

FAO (2012), *Estado de las Áreas Marinas y Costeras Protegidas en América Latina.* Available at, www.fao.org/3/a-as176s.pdf.

FAO (2018), *El Estado Mundial de la Pesca y la Acuicultura. Cumplir los Objetivos de Desarrollo Sostenible.* Available at, www.fao.org/publications/sofia/es/.

García, R. (2018), *Pescadores Protestan contra Concesiones Petroleras en el Norte del Perú.* Available at, https://es.mongabay.com/2018/04/pescadores-paro-petroleo-oceanos-peru/.

Grupo Propuesta Ciudadana (2014). *Concesiones Mineras en el Perú. Analisis y Propuesta de Políticas.* Available at, http://redextractivas.org/concesiones-mineras-peru-analisis-propuestas-politica-grupo-propuesta-ciudadana-2014/.

Hooker Mantilla, Y. (2011), Criterios para la Zonificación de Áreas Marinas Protegidas en el Perú. Unidad Marino Costera, Servicio Nacional de Áreas Naturales Protegidas por el Estado. *Revista Areas Marinas Protegidas – Perú,* 1. Available at, http://old.sernanp.gob.pe/sernanp/archivos/biblioteca/publicaciones/islas/revista1.pdf.

IMARPE (2010), *Informe Nacional sobre el Estado del Ambiente Marino del Perú.* Available at, http://cpps.dyndns.info/cpps-docs-web/planaccion/docs2010/oct/XVII_AG_GC/18.Contaminacion.marina.Informe.final.Peru.pdf.

Instituto Nacional de Estadística e Informática (INEI) (2017), *Censos Nacionales 2017: XII de Población, VII de Vivienda, III de Comunidades Indígenas.*

Instituto Nacional de Estadística e Informática (INEI) (2012), *I Censo Nacional de la Pesca Artesanal Ámbito Marítimo.*

Maguiña, A. (2009), Entre el Mar y la Arena: Las Zonas Marino-Costeras. *Foro Jurídico 9*. Available at, http://revistas.pucp.edu.pe/index.php/forojuridico/article/viewFile/18534/18774.

MINAM (2012), *Los Desafíos para la Zona Costera Peruana en el Siglo 21*.

MINAM (2015), *Estudio de Desempeño Ambiental 2003–2013*. Available at, https://sinia.minam.gob.pe/contenido/conoce-estudio-desempeno-ambiental-esda-peru-2003-2013.

MINAM (2016), *La Lucha por la Legalidad en la Actividad Minera (2011–2016). Avances Concretos y Retos para Enfrentar la Problemática de la Minería Ilegal y Lograr la Formalización de los Operadores Mineros*. Available at, www.minam.gob.pe/informessectoriales/wp-content/uploads/sites/112/2016/02/12-La-lucha-por-la-legalidad-en-la-actividad-minera.pdf.

MINEM (2017), Dirección de Promoción Minera. *Anuario Minero 2017*. Available at, www.gob.pe/institucion/minem/informes-publicaciones/112024-anuario-minero-2017.

Ministerio del Medio Ambiente de Chile (2014), *Quinto Informe Nacional de Biodiversidad de Chile*. Available at, https://mma.gob.cl/wp-content/uploads/2017/08/Libro_Convenio_sobre_diversidad_Biologica.pdf.

OCEANA (2016), *Áreas Marinas Protegidas: Recomendaciones para su Gestión y Aprovechamiento Sostenible*.

OEFA (2018), *OEFA en Cifras al II Trimestre – 2018. Reporte Estadístico*.

ONU México (2016). *Mexico y su Biodiversidad*. Available at, www.onu.org.mx/diversidad-biologica/ (accessed December 11, 2018).

PERUPETRO (2019). *Mapa de Lotes*. Available at, www.perupetro.com.pe/ (accessed January 5, 2019).

Sierra Praeli, Y. (2018), *Perú: ¿Conservación Marina versus Actividad Petrolera?* Available at, http://bloglemu.blogspot.com/2018/04/peru-conservacion-marina-versus.html.

Sistema Nacional de Información Ambiental (SINIA) (2017), *Cifras Ambientales 2017*.

Sociedad Nacional de Pesquería (2017a), *Industria Pesquera: Contribución a la Economía Peruana*.

Sociedad Nacional de Pesquería Revista Institucional de la Sociedad Nacional de Pesquería. Año XIX, Edición 102 (2017b), *Sector Pesquero: Motor de la Economía*.

Subsecretaria de Pesca y Acuicultura del Gobierno de Chile (2012), *Estado de Situación de las Principales Pesquerias Chilenas, Año 2017*. Available at, www.subpesca.cl/portal/618/articles-100052_recurso_1.pdf.

UNEP – World Conservation Monitoring Centre. (2011), *Marine and Coastal Ecosystem Services. Valuation Methods and their Practical Application*.

6 Marine protected areas

Pedro Solano and Alfredo Gálvez

Introduction

Natural protected areas have historically shown to be the most efficient and recognized instruments to ensure the protection of biological diversity, the generation of knowledge and awareness among citizens regarding the value of the natural patrimony. Nearly 150 years have gone by since the establishment of the first continental National Park in the world[1] and 116 years since the first marine protected area (MPA) was established.[2] The overall balance is very positive regarding their contribution to global conservation. Thanks to natural protected areas, and as a contribution to a true human heritage, areas such as Patagonia, Antarctica, Galapagos Islands or a large part of the sea and Polynesian Islands, as well as huge extensions of the Amazon and African savannah have persisted almost in their natural state for decades, to the astonishment, pride and enjoyment of generation after generation. Natural protected areas remind us that we are a planet of living entities, diverse and unique ecosystems, incomparable places of beauty, valuable natural resources for living and of great interdependence among the biological world, including human beings.

Although natural protected areas have become widely recognized, are well regarded and have consolidated as proven conservation tools, they continue to be undervalued and have become highly vulnerable and threatened spaces. Biological diversity itself is still, to a large extent, a mystery for humanity despite the research undertaken for centuries. The knowledge about biological diversity in general, whilst considerable, is still quite basic and fragmented. This is particularly clear in the case of the sea and oceans which some argue are less known than the moon itself! With regards to the seas and oceans, gaps in information and knowledge are very great when contrasted with what we know about continental biological diversity. Whereas the planet's sea (water) surface is vastly superior to the earth's surface – three times bigger – protected extensions in the sea and ocean are ten times smaller than those on land. Only 7.4 percent of the ocean is subject to legal protection under the notion of MPAs even though threats and conservation needs are massive.[3]

In 2010, the Conference of the Parties to the Convention on Biological Diversity (CBD) agreed on the Biodiversity or Aichi Targets by which parties commit to protect by 2020 at least 17 percent of terrestrial and inland water, and 10 percent of coastal and marine areas. In 2015, this same target was incorporated as one of the United Nations Sustainable Development Goals.

The challenge to protect and manage marine spaces efficiently is enormous and at the same time urgent. It is becoming increasingly evident that in the context of global climate change, social and economic models will change drastically in this century. Having healthy oceans and marine environments contributes to food security, generates greater resilience, facilitates transportation and provides huge recreational opportunities. In this particular scenario, the importance of an MPA will increase as a means to ensure well-functioning oceans. This chapter presents some ideas and reflections regarding the role and importance of an MPA.

Concepts and evolution

The World Conservation Union (IUCN) defines a marine protected area or MPA in general as any area of intertidal or subtidal terrain, together with its overlying water and associated flora, fauna, historical and cultural features, which has been reserved by law or other effective means to protect part or all of the enclosed environment (Kelleher, 1999). This must be read jointly with the IUCN definition for natural protected areas which refers to a clearly defined geographical space, recognized, dedicated and managed, through legal or other effective means, to achieve the long-term conservation of nature with associated ecosystem services and cultural values. The Technical Expert Group on Marine and Coastal Protected Areas of the CBD on the other hand, defines MPAs as

> any confined area within or adjacent to the marine environment, together with its overlying waters and associated flora, fauna, and historical and cultural features, which has been reserved by legislation or other effective means, including custom, with the effect that its marine and/or coastal biodiversity enjoys a higher level of protection than its surroundings.
>
> (SCBD, 2004)

Protected areas are legal forms or tools for biodiversity conservation and can be classified according to certain criteria. This will depend on factors including: their conservation objectives or targets; the legal regime under which they are created; their governance and management models, among others. This classification allows states to select among different types of governance and management categories. Although all protected areas seek conservation objectives for the public interest, not all protected areas are the same. There are different types of governance and management categories. The categorization and creation of marine protected areas is usually approved through a legal norm (i.e. a law, regulation, resolution, etc.). In this regard, a first approach to classify and categorize MPAs derives from a practice where, during the early years, the concepts of marine reserves, marine parks and marine sanctuaries were used.[4]

Following this, IUCN set an internationally recognized classification, to create a certain uniformity and allow these spaces to "fit" within a comparative and recognizable framework. Marine protected areas originally focused on specific species and ecosystems (see Box 6.1). To achieve this, IUCN proposed six different categories of natural protected areas. Even though this classification system is not mandatory or binding for states it has been broadly and extensively applied. Its key principles include: setting primary objectives for each area; management models; levels of human intervention and sites' usefulness, representativeness and characteristics (Lausche, 2012).

It is worth noting that although many of these categories were thought out for terrestrial spaces, IUCN has attempted to describe them in a functional way, that can be adapted to marine environments. In accordance with the "Guidelines for Applying Protected Areas Management Categories" published in 2008 by IUCN (Dudley, 2008) the current system includes six categories:

I   Strict nature reserve/wilderness area
II   National park
III   Natural monument or feature
IV   Habitat/species management site
V   Protected landscape/seascape
VI   Protected area with sustainable use of natural resources

This same publication establishes principles for the application of IUCN categories to MPAs. One interesting aspect is "vertical zoning" for marine areas in contrast to more horizontal zoning which applies primarily to land areas. Both a horizontal and vertical zoning would allow for more accurate estimates of the extension of the marine area. Clearly, the complexity of marine areas is quite different to that of terrestrial areas.

A third classification, outside the MPA framework itself, but with an impact on the potential conservation of marine biological diversity, derives from jurisdictional aspects and their connection to international agreements and institutions. For instance, a reference could be made to marine areas beyond jurisdictional waters and the role of Specially Protected Areas under the Convention for the Protection of the Marine Environment of the North-East Atlantic (OSPAR) or to Whale Sanctuaries under the competence of the International Whaling Commission (IWC) or to Particularly Sea-Sensitive Areas as defined by the International Maritime Organization (IMO) pursuant to the International Convention for the Prevention of Pollution from Ships (MARPOL). In practice, what will make a difference among these different categories is their overall purpose and protection requirements but given their transboundary nature, implementation is more complex and subject mainly to international law. The global coverage of MPAs is estimated at around 7.47 percent of oceans. This means 27,061,861 km^2 and 14,841 legal marine protected areas around the world.

Box 6.1 The first marine protected areas

The first marine protected areas derived from mechanisms used for the protection of coastal and marine species such as pelicans and guano birds, and from the identification of natural monuments or special sites such as coral reefs. Below are several examples:

Pelican Island in the United States was declared a refuge in 1903, with the goal of protecting sea birds, mainly the brown pelican (*Pelecanus occidentalis carolinensis*).

Glacier Bay National Monument in Alaska was established in 1925 over an extension of 1,135,000 hectares. However, the effective protection of the adjacent marine space is unclear. The main conservation objective is the coastal–marine glaciers.

Fort Jefferson National Monument in Key West, Florida, United States, established in 1935 with a sea extension of 18,850 hectares, leaving a small terrestrial portion to protect the coastal fort constructed following the American Civil War.

Green Island was established in Australia in 1938, a part of the Great Barrier Reef.

With regard to Latin America, Peru has provided legal protection to the **Guano Islands** and **Capes** and adjacent sea since the end of the nineteenth century. Since the twentieth century there have been legal norms that ban fishing and other activities surrounding these sites, as well as disrupting the habitat for guano-producing birds. After 150 years of "soft" legal protection, in 2009 Peru declared 22 guano islands and 11 capes as protected areas.

Guadalupe Island in Mexico was established in 1922, as well as the **Chacahua Lagoons**, declared in 1937. Although these two areas have enjoyed more of a coastal management approach rather than a marine-based approach, during the last three or four years the focus has shifted towards the inclusion of marine and oceanic conservation strategies.

In the 1970s, Chile declared the first Natural Sanctuaries under the Law on National Monuments whose objective was to protect in a restrictive manner marine–coastal spaces, as well as zones of special scientific relevance. Marine reserves in coastal zones have been established such as the **Humboldt Penguin National Reserve**.

International context and global targets: Aichi Biodiversity Targets of the Convention on Biological Diversity and United Nations Sustainable Development Goals

The Tenth Meeting of the Conference of the Parties to the Convention on Biological Diversity (CBD), held in Japan in 2010, proposed a new approach for member countries to adapt their National Biodiversity Strategies. This approach incorporated what is known as the "Strategic Plan for Biological Diversity 2011–2010 and the Aichi Biodiversity Targets," and includes 5 strategic goals and 20 world targets. Within these world targets is Target 11, related to the effective protection of at least 10 percent of Member States' marine waters, through conservation tools (i.e. natural protected areas and other efficient conservation measures) that are effective and equitable in management practices, well represent environments and eco-systems, stimulate systematic connectivity and integration of the landscape, and act as effective area-based measures – whether or not classic protected areas per se.

Since the Aichi targets "expire" in 2020, a strong trend to establish marine protected areas has been seen. Not only are seas and oceans protected but countries also exceed in meeting their targets. This has been the case in the recognition of large sea extensions in countries such as Chile, Ecuador, Colombia, Costa Rica, Mexico and the United States, among others. This also shows how international law can play a trend-setting role and encourage and accelerate global responses towards common objectives.

In line with the Aichi Targets, during the United Nations Conference on Sustainable Development Rio+20, governments reached a decision to design and approve a set of concrete objectives so that their actions, policies and laws/regulations are on course towards sustainable development. These 17 goals are the "Sustainable Development Goals" (SDG), some of which are strongly related, either directly or indirectly, to marine protected areas. These include SDG 14 (life below water); SDG 1 (no poverty); SDG 2 (zero hunger); among others. The United Nations SDGs are also in line with reaching at least 10 percent of protected sea and ocean surface.

Comparative advances on marine protected areas in Mexico, Chile and Peru

Mexico

Mexico has 182 federal natural protected areas system, of which 37 are found in marine and coastal ecosystems (70,212,782.04 hectares).[5] This makes Mexico a leader in the region on marine conservation. These and land spaces jointly make up more than 90 million hectares and are administrated by the National Commission of Natural Protected Areas which depends on the Secretariat of Environment and Natural Resources (SEMARNAT).

Mexico established its first marine protected area in 1922, Guadaloupe Island,[6] becoming a pioneer in terms of marine conservation in the region. Later, towards 1937, the first marine and coastal area of the Chacahua Lagoons was created and recognized as a National Park, located in the State of Oaxaca. Currently, 7 out of 17 Mexican states have established coastal natural protected areas and 4 of them have included sea surfaces as part of these areas. The Federal State has competence over these marine zones.

The General Law of Ecological Equilibrium and Environmental Protection and its regulation and the General Wildlife Law govern the Mexican system of protected areas, including by defining the management regime applicable and permitted uses of resources and activities. Out of six legal categories of protected areas recognized in these laws, four have been used to protect marine and coastal spaces: Biosphere Reserves, National Parks, Flora and Fauna Protection Areas and Sanctuaries (FAO, 2012). In terms of marine and sea areas in general, the Federal Law of the Sea grants the Marine Secretariat control and monitoring powers over these spaces and the Fishery Law governs fishing activities inside and outside MPAs. Likewise, the General Law of Ecological Equilibrium and Environmental Protection defines general competences regarding the establishment, administration and monitoring of MPAs, to be conducted in a coordinated manner by the National Commission of Natural Protected Areas and the Marine Secretariat. As in many other places in Latin America, the Mexican Navy plays an important role in monitoring and patrol, as well as providing navigation authorizations. Competences and jurisdiction among government agencies in charge of protected areas are relatively well delimited and are reflected in most management plans of MPAs.

Chile

In recent years, Chile has given a significant push to the creation of marine protected areas and just like Mexico, has become a leader country in marine conservation. This is reflected in terms of the protection of natural spaces and also culturally relevant areas for indigenous peoples as in the case of Rapa Nui of Easter Island)[7] and different traditional uses concerning hydrobiological resources. Chile has 33 MPAs that represent more than 3 million km^2 and includes recent decisions on the creation of new Marine Parks and Marine Reserves, such as Parque Marino Nazca-Desventuradas.[8]

Chilean legislation recognizes four MPA categories[9]: Marine Parks and Marine Reserves that protect specific ecosystems and Nature Sanctuaries and Marine Protected Areas of Multiple Use that cover marine and/or coastal spaces. In the case of Marine Parks, they are strictly protected important ecological units for science, under a condition of indirect use. Marine Reserves, on the other hand, specifically protect hydrobiological resources and their breeding and restocking zones, under a direct use approach by phases, prior to approval by the Sub-Secretariat for Fisheries and Aquaculture. The management of these areas rests on the National Forest Corporation, which has both land and marine competences.

The mandate for the management of marine and coastal protected areas is in Law 19300, the General Environment Law and Law 18362 that establishes the National System of Wilderness Protected Areas. Under these laws, the state must administrate the National System, which includes Marine Parks and Reserves. Competences regarding marine and coastal areas are shared. Many ministries have a responsibility for regulating, granting a series of rights and undertaking inspections of different activities within or close to marine areas. For instance, under Law 18892, the General Law on Fisheries and Aquaculture, the Sub-Secretariat of Fisheries is responsible for fishery activities or uses of hydrobiological resources in marine protected areas.[10] Although decision-making becomes complex with this distribution of different competences among different agencies for different aspects of MPAs, Chile has made considerable progress in marine conservation in the region. A clear example has been to increase the percentage of protected sea in less than two years (2016–2018), from 13.4 percent to 42 percent with the creation of MPAs Rapa Nui, Juan Fernández and Cabo Esperanza.[11]

Peru

In the case of Peru, marine conservation takes place in a context where the sea has been viewed, for decades, as a purely extractive source of wealth. Two examples stand out: guano, taken from the Peruvian islands since the nineteenth century, was used as a potent natural fertilizer in the development of intensive agriculture and the boom of fishery industries in the 1970s made Peru one of the leading fishing nations in the world.

Numerous research expeditions and activities have taken place looking at populations of guano birds off the Peruvian coast, through the Institute of the Sea of Peru and other research organizations and universities. Considerable data and information on historical patterns regarding extraction of guano, "anchovy" and other natural resources have been produced (IMARPE, 1966).

The case of guano is particularly interesting, given that at one time it was the main source of income for the Peruvian state. It is estimated that in 1861, 80 percent of the general budget was covered by contributions generated by the guano economy. This explains the legal protection given to marine and terrestrial spaces, as well as the species that generated them. These areas have been crucial to protect other species and their habitats all along the Peruvian coast.[12]

Apart from the guano story, in 1975, Peru started the creation of marine and coastal protected areas. The Paracas National Reserve covers 65 percent of marine waters (217,594 hectares) within its extension, with the main objective of ensuring the preservation of a unique portion of the Peruvian marine-coastal ecosystem influenced by the Humboldt or Peruvian Current.[13] This National Reserve has become an emblematic area for marine and coastal conservation. Its conservation and management goals have mostly responded well to the accommodation of aquaculture management tools, including special concessions for fish farms.[14]

The Peruvian National System for Protected Areas was gradually strengthened during the 1990s. The National Institute for Natural Resources of the Ministry of Agriculture was responsible for management of and developing policy on protected areas in general. It wasn't until 2008 that with the creation of the Ministry of Environment, a specialized governing body for the National System was created: the National Service of Natural Protected Areas. This led to a completely new approach to protected areas management and policy making, including through a renewed interest in *marine* protected areas.

In 2009, the National Reserve System for Guano Islands, Islets and Capes was created, covering most areas originally protected for guano production. The Reserve System extends over 140,833 hectares. Almost in parallel, the Illescas Reserved Zone (in Northern Peru) was established in 2010, with 37,452 hectares and in 2011, the San Fernando Reserve Zone with 154,716 hectares was recognized as a National Reserve – the first step towards its recognition as a protected area.

Even with these advances, and the creation of these protected areas and a National Reserve, protected marine ecosystems in Peru are very much underrepresented, covering just over 0.60 percent of the country's marine jurisdiction, nowhere near Aichi Target 11. All of Peru's current protected areas are within the influence of the cold Humboldt Current. There are plans and a drive (mostly from civil society) to create a tropical marine protected area in Northern Peru but political will and commitment is still lacking for this initiative. It should be said, however, that Peru has not signed the Convention on the Law of the Seas and considers sovereignty over 200 miles of territorial sea which furthermore depresses the percentage.

Protected areas are governed mainly by Law 26834, the Law of Natural Protected Areas, and its regulation, approved by Supreme Decree No. 038–2001-AG. Under this legal framework, nine categories for protected areas are recognized – three for indirect use and six for direct use. Only one of these categories – National Reserve – has been used to protect marine and coastal spaces. This is a direct or sustainable management use category with sufficiently flexible characteristics that make sustainable use and extraction of natural resources (i.e. fish) possible, in accordance with area management tools.[15]

Box 6.2 offers a brief comparative look at Mexico, Chile and Peru in terms of recently established MPAs and sea/ocean protection measures.

Challenges for the management of marine protected areas

Fisheries and other extractive activities

Many marine protected areas established in recent years are located within the 5-mile zone near the coastal border. These zones usually have a high

Box 6.2 Protection of marine areas in Chile, Mexico and Peru

Country	MPA/extension/year created	Percentage of the jurisdictional sea protected
Chile	– Mar de Juan Fernández, Diego Ramírez, Islas Rapa Nui and Cabo de Hornos (463,000 km²) – 2017- Nazca-Desventuradas (300,000 km²) – 2016- 5 new MPAs in Patagonia (40,500 km²) – 2018	42%(Higher than the percentage protected of continental Chile)
Mexico	– Reserva Marina de Revillagigedo (148,000 km²) – 2017	22.3%
Peru	Reserva Nacional de San Fernando (154,716.37 km²) – 2011	0.60%

Source: own elaboration by the authors.

presence of local fishers, which creates tensions. Apart from having to protect habitats, species, landscapes and ecosystems, MPAs in these zones must guarantee sustainable fishing and extraction of hydrobiological resources by artisanal fishers who have been accessing these zones for generations. On the other hand, discussions on hydrocarbon and mineral extractions in MPAs continue and in many cases are paralyzed or the processes to establish MPAs have been reversed. Some of these activities, with correct technologies and safeguards in place, may not be completely incompatible with the conservation of ecosystems and natural resources. Appropriate management of marine spaces requires commitment by all and the need to approach the hydrocarbon and mining sector, as well as the fishers' unions, in order to advance with a joint agenda that accelerates marine conservation.

No-take zones

"No-take zones" are areas where renewable and non-renewable resources extraction is prohibited. These areas have the best guarantee for restocking and to compare zones which have had interventions with those that have not. Marine protected areas complemented with "no-take zones" act as "long-term deposits" for the future of mankind. However, this strict management tool is rarely used, and when it is, it applies over areas for where there is no economic or social interest or activities, often not covering the most valuable places for biodiversity conservation.

Institutional aspects

Much remains to be done in terms of capacity-building and tools for MPA management authorities. In most countries, the protected areas systems were designed with a terrestrial and land-based perspective. The imbalance of knowledge, understanding and management capacity for marine protected areas is considerable compared to their land protected areas. Converging and often overlapping competences in marine areas due to either fishery of marine navigation and transport issues, reduce the possibilities of appropriate management units.

Marine protected areas networks and corridors

Another important challenge for marine biodiversity conservation in the region relates to limited national and international networks for marine protected areas. These networks often serve to trigger "biological corridor" initiatives which are key for the protection of marine species such as mammals and sharks. These tools are critical in ensuring appropriate combat against ecosystem degradation and loss of key species, as well as in reflecting commitment to international conservation goals. A good example of collaboration and increased action is the Galapagos (Ecuador)–Malpelo (Colombia)–Isla Cocos (Costa Rica) corridor, covering an extent of 1,383,643 km^2. In the case of Peru and Chile, the possibility of protecting an extensive and shared marine ecosystem may not be that far off, as could be the case of the Cordillera and Nazca seamounts (Gálvez-Larach, 2009).[16]

Monitoring and control

Monitoring and controlling marine protected areas is extremely complex, not only because of the practicalities of covering large areas, but also due to the multiplicity of actors and state entities present in such spaces. This includes overlaps in competences according to specific resources, resulting many times in a lack of supervision for illegal threats and activities. This is a major challenge which requires careful review of legal and institutional frameworks, and a desire to sometimes gain and often lose certain competences, or at the very least create appropriate and effective coordination procedures. Likewise, creating limits and conditions with regards to MPAs, and its effect on users and local actors is often not easy to achieve. Participation and continued interaction between state and local actors is key to create enabling conditions to establish necessary use restrictions and limits within certain boundaries and areas traditionally accessed by communities. This can furthermore enhance possibilities of legitimate future control, supervision and monitoring of these spaces. Technological advances would certainly be a decisive factor to improve this and support monitoring of all marine activities within a particular area in real time.

Compliance to international commitments

The Aichi Targets and slightly more recent Sustainable Development Goals are becoming determinant factors to increase the protection of marine ecosystems. Although it is unlikely that the Aichi Targets will be met fully, the reality is that without them there would have been no further progress on these issues in recent years. Within a short period of time, compliance can be reviewed and new and ambitious targets set, with all the emerging additional knowledge on the role of the oceans to face climate change and food security made available. Specific targets on the disposal of wastes at sea and above all the pollution of oceans and impacts on species and ecosystems from plastics will also be high on the international and national agendas for years to come.

Notes

1 Yellowstone, 1872.
2 Pelican Island, 1903.
3 www.protectedplanet.net/marine.
4 Legislation in the UK, New Zealand, Australia and India has completed and refined these categories.
5 See, https://simec.conanp.gob.mx/aichi/Mexico_Meta_11_Dic_2016_Cuadernillo. pdf.
6 Source: TNC, Rivera-Arriaga *et al.*, 2004.
7 Under the law in Chile, Marine and Coastal Areas for Indigenous Peoples can be created. The state grants the management of a certain marine space to an association of communities which must present an Administration Plan prior to this recognition.
8 The Marine Nazca-Desventuradas Park covers over $300,000\,km^2$.
9 Law 18,892, General Law on Fisheries and Aquaculture recognizes Marine Parks and Marine Reserves as distinct categories.
10 The General Law on Fisheries and Aquaculture in Chile recognizes two categories of MPAs where the Sub-Secretariat of Fishery has competences regarding recognition of these areas and management of resources therein.
11 Source: www.unenvironment.org/news-and-stories/story/latin-american-and-caribbean-countries-champion-marine-conservation.
12 For further information see, Solano, 2005, pp.243–250.
13 Supreme Decree No. 1281–75-AG.
14 This legal figure derives from the Regulation for the Management of Special Concessions for Benthonic Species in the Paracas National Reserve approved by Supreme Decree No. 023–2001-PE, which in turn responds to the regulation on protected areas regarding hydrobiological resources.
15 Master plan for protected areas, management plan for concessions, etc.
16 In 2017 Chile and Peru signed an agreement for the conservation of marine and coastal areas. See, http://portal.mma.gob.cl/chile-y-peru-firman-acuerdo-sobre-conservacion-de-areas-protegidas-marino-costeras/.

References

Castilla, Juan Carlos. (1996). La futura Red Chilena de Parques y Reservas Marinas y los conceptos de conservación, preservación y manejo en la legislación nacional. *Revista Chilena de Historia Natural* 69:253–270.
Dudley, N. (2008), *Directrices para la Aplicación de las Categorías de Gestión de Areas Protegidas.* UICN. Gland, Switzerland.

Elbers, J. (ed.) (2011). *Las Áreas Protegidas de América Latina: Situación Actual y Perspectivas para el Futuro.* Quito, Ecuador, UICN.

FAO (2012). *Estado de las Areas Marinas y Costeras Protegidas en América Latina.* Written by Aylem Hernández Avila. REDPARQUES Cuba. Santiago de Chile.

Gálvez Larach, M. (2009). Montes submarinos de Nazca y Salas y Gomez: una revision para el manejo y conservación. *Latin American Journal of Aquatic Research* 37(3): 479–500.

IMARPE No. 10. (1966) Available at http://biblioimarpe.imarpe.gob.pe/bitstream/ 123456789/239/1/INF%2010.pdf.

Kelleher, G. (1999) *Guidelines for Marine Protected Areas.* IUCN Publication. Gland, Switzerland.

Kelleher, G. and Kenchington, R. (1992). *Guidelines for Establishing Marine Protected Areas.* A Marine Conservation and Development Report. IUCN, Gland, Switzerland.

Lausche, B. (2012), *Directrices para la Legislación Relativa a Areas Protegidas.* UICN, Gland, Switzerland.

Rivera-Arriaga, E., Villalobos-Zapata, G.J., Azuz-Adeath, I., Rosado, F. (eds) (2004). *El Manejo Costero en México. Capítulo 14 – Las Áreas Naturales Protegidas Costeras y Marinas de México.* Universidad Autónoma de Campeche, SEMARNAT, CETYS-Universidad de Quintana Roo, – May, pp. 191–222.

Secretariat of the Convention on Biological Diversity (SCBD). (2004), Technical Advice on the Establishment and Management of a National System of Marine and Coastal Protected Areas. CBD Technical Series no.13, SCBD.

Sierralta L., Serrano, R., Rovira, J. and Cortés, C. (eds) (2011), *Las Areas Protegidas de Chile.* Ministerio del Medio Ambiente. Santiago de Chile.

Solano, P. (2005), *La Esperanza es Verde.* Sociedad Peruana de Derecho Ambiental – SPDA. Lima, Perú.

7 Marine and coastal planning

Fernando A. Rosete Vergés

Introduction

As a result of intense and ongoing degradation of oceanic ecosystems, marine and coastal planning has become especially relevant in recent years as a tool to confront environmental problems.[1] Marine and coastal environment degradation is clearly manifested in different regions and is caused, mostly, by direct and indirect human activities. It is therefore critical to undertake planning activities for both marine and coastal zones to reduce human-derived impacts on ecosystems and find ways to balance ecosystem conservation and the harmonization of natural processes with pressing needs of human populations (Botero *et al.*, 2016).

It is worth noting that to a great extent, under neoliberal economic models, tools and policies to prevent the collapse of ecosystems and societies in general, are limited or non-existent (WWF, 2018). Global environmental degradation processes are so intense, that in the case of global warming for instance, even if greenhouse gas emissions were cut altogether, its impacts would be permanent and grave for years to come (IPCC, 2018). This situation forces us to rethink the way in which natural resources are exploited and how this has consequences on marine and coastal ecosystems in particular.

This chapter reviews the main international and regional instruments available which support marine and coastal planning. To understand the importance of these instruments for conservation and sustainable management of ecosystems and their resources, the chapter has been structured as follows: first, a brief description is offered about the present marine and coastal environmental situation globally which justifies planning for the conservation and management of marine and coastal ecosystems; second, a review is made of the most widely used instruments of marine spatial planning and how this is reflected in examples around the world and in the Latin American region; finally, the conclusion, and recommendations are made on how to improve and strengthen planning instruments and processes.

The general marine and coastal environmental situation

Concern for the health of marine and coastal ecosystems has been expressed forcefully in recent decades and their importance acknowledged in all international United Nations sponsored meetings.

As a result of the United Nations Conference on Sustainable Development – Rio+20 (2012), oceans and marine ecosystems were recognized as critically important in the context of conservation and sustainable development efforts to respond to the threats of pollution and climate change (ONU, 2012). The related 2030 Agenda and its Sustainable Development Goals (ONU, 2016) also recognize the importance of reducing impacts on marine ecosystems and their resources. Goal 14 acknowledges that oceans are vital for human survival and the planets viability, thus careful management of ecosystems and their resources is key to reach a sustainable future. And so on with numerous governmental and non-governmental processes worldwide.

However, regardless of global concerns, according to the Heinrich Böll Foundation "Ocean Atlas 2017" (HBS, 2017), the planet currently faces an oceanic crisis due to growing stresses imposed on its ecosystems. As a result of global climate change, oceans and seas are suffering from acidification, warming and rising levels, processes that have significant repercussions on communities of marine organisms and coastal zones. Oceans are largely used as inexhaustible dumps and currently receive more waste than they can processed naturally. Humans extract more than what oceans can replenish – 90 percent of fisheries around the world are at their maximum limit, or have already exceeded extractions beyond the ocean's regeneration capacity (Pauly *et al.*, 2002). The demand for non-renewable natural resources such as minerals and energy from the deep-sea bed is high and will continue to rise in the future, with the risk that deep sea mining will destroy complete ecosystems before we even know it.

One of the most emblematic impacts is the accumulation of plastic waste in the oceans. At present, five huge islands with high levels of plastic pollution per km^2 have been "created" due to the swirls of surface currents in the North and South Pacific, North and South Atlantic and Indian Ocean. Out of the five, the largest and with greater quantities of waste are located in the Northern Hemisphere and South Atlantic (UNEP and GRID-Arendal, 2016). Clearly this is not just a matter of waste production and poor management in coastal zones but, rather, a clear example of the impact of activities conducted inland that end up impacting seas and coasts.

Similarly, "dead zones" from the lack of oxygen in the Gulf of Mexico and the Baltic Sea have been caused by excessive nutrient pollution mainly from land agricultural activities (Douvere and Ehler, 2011). In the case of the Gulf of Mexico, "dead zones" are produced during the summer, caused by nitrates being discharged into coastal waters from the Mississippi River as a result of fertilizers used in intensive agriculture and by hog farms. Agricultural and industrial activities undertaken in the United States all the way from the Canadian border southwards, end up impacting the Gulf of Mexico. This highlights the need for a river basin approach in the efforts to control negative impacts of land-based activities on marine and coastal zones (HBS, 2017).

Coral reefs are marine ecosystems that are especially vulnerable to alterations caused in the seas and oceans. Their biological and economic importance has been widely documented (Spalding *et al.*, 2001; Carballo *et al.*, 2010; Kritzer *et al.*, 2016; Pondella *et al.*, 2016; Calderón-Aguilera *et al.*, 2017; Veron *et al.*, 2018). However, coral reefs remain in a critical situation and a worst-case scenario predicts their total collapse in the coming decades (Donner *et al.*, 2008; Baskett *et al.*, 2009; Teneva *et al.*, 2011).

In the case of coastal areas, the situation is not much different. The pace and intensity of change is on the rise given they are the most dynamic geographical areas on the planet due to their interaction with land, sea, the atmosphere, as well as human activities. An important percentage of the world's population lives in coastal areas.[2] They are preferred areas for the tourist industry and many economic and strategic activities, such as the extraction and transformation of energy (Ansong *et al.*, 2017; Muñoz and Le Bail, 2017).

The importance of coastal and marine planning and management

Planning of human activities in a confined territory dates back to the ancient civilizations. Since the beginning of the eighteenth century, planning has been a key tool to geographically balance populations, organize production and create wealth for societies (Rosete and Fuentes, 2018).

More recently, marine coastal planning has gained traction since the Rio de Janeiro Earth Summit in 1992, where the recognition of Integrated Coastal Zone Management (ICZM) gave way to planning initiatives to face challenges in coastal and marine areas (Botero *et al.*, 2016). Among the most frequent problems addressed by coastal and marine planning is the recovery of important commercial fishery species, restoration of ecosystems, preservation of environmental services and control of inland pollution and sediment discharges (Long *et al.*, 2015; Ansong *et al.*, 2017).

Although many specific coastal management programs with a focus on ICZM have been developed and implemented, few have met the objectives and goals initially set. This is due to difficulties in coordination among different levels of government and actors with different interests and stakes in these areas (Domínguez-Tejo *et al.*, 2016; Buhl-Mortensen *et al.*, 2017; Rodriguez, 2017; Nava *et al.*, 2018).

Even with these shortcomings, there are examples where environmental problems have been effectively addressed (Pomeroy *et al.*, 2014).[3] Furthermore, when there is political determination and sensitivity and awareness from actors involved, results can be very promising (Ansong *et al.*, 2017).

Instruments for coastal and marine planning

The main marine planning instrument or tool is Marine Spatial Planning (MSP). According to Ehler and Douvere (2009) MSP can be defined as a public process for analyzing and allocating spatial and temporal distribution of human activities in marine areas to achieve ecological, economic and social objectives that are usually specified through a political process. MSP is comprehensive and multi-objective, strategic, with a forward-looking approach, continuous, adaptive, participatory, eco-systemic and applied (Ehler and Douvere, 2009).

Comprehensive because it should include all economic sectors; multi-objective because it should incorporate social, ecological and economic objectives; continuous given it is a process that includes monitoring and follow-up of implementation; evaluated based on learning, and therefore adaptive to new existing conditions of a new cycle; participatory as actors should engage actively in the elaboration, management and implementation process; eco-systemic as it seeks to maintain ecosystems and their interactions over time; and applied as it focuses on marine spaces occupied by the people who are also active players in their care (Iglesias-Campos *et al.*, 2014). Nevertheless, in terms of development of MSP, various approaches commonly used in planning can be adopted. These include: watershed, ecosystem-based management, coastal zone integrated management, integrated management of coastal and marine resources or sustainable land management approaches (Pomeroy *et al.*, 2014). In 2009, the Intergovernmental Oceanographic Commission (IOC) and UNESCO released a guide to elaborate MSP with a focus on ecosystem-based management approaches.[4]

Relevant experiences at the international level

As of August 2018, the UNESCO Marine Spatial Planning Program had 140 registered plans in 70 countries on five continents, with different levels of progress and advances. Seven indicators of progress have been identified (pre-planning, planning analysis, plan development, finished plan, approved plan, plan implementation and review of the plan) in the MSP process (UNESCO, 2018a).

In Africa, only the Seychelles has a finished plan for its exclusive economic zone (EEZ). In Asia, China has implemented 12 plans, one for its EEZ and 11 provincial plans. The Philippines is also implementing a plan in the province of Bataan, as well as Vietnam in the Municipality of Danang. In the case of Oceania, Australia has implemented and even revised the plan for the Great Barrier Reef Marine Park, as well as developed and completed five regional plans. Kiribati is implementing another plan for the protected area of the Phoenix Islands. New Zealand has a finished plan for the Hauraki Gulf. There are four countries in Europe which have made significant progress, having reached a review phase: Belgium for its EEZ, Germany with the Mecklenburg-Vorpommern regional plan, Holland with the territorial sea and the EEZ and Norway with the plan for the Barents Sea and Lofoten Islands. Various other plans are being implemented: one in Croatia (Zadar County), four in Germany (EEZ of the Baltic Sea, EEZ of the North Sea, the Schleswig-Holstein region and Niedersachsen region), and two in Norway (Norwegian Sea and the North Sea). Great Britain has 12 plans, two formally approved for their implementation. Box 7.1 briefly describes experiences with advances in marine planning implementation processes.

Box 7.1 Relevant experiences in the MSP process

Great Barrier Reef Australia

The expertise created as a result of the adaptive management of the Great Barrier Reef Marine Park is the most recognized international milestone for MSP. Three key components have contributed to its success: collaboration between the Australian government and the Province of Queensland; use of available scientific information to inform the process; and enabling aboriginal peoples and Islanders of the Torres Strait to lead and conduct the process and be recognized as the traditional owners of the Great Barrier Reef and its derived culture which continuously connects the land and sea.

The MSP process was initiated when the Marine Park was established in 1975. In 1979 a management program was implemented to address land-based actions impacting the Great Barrier Reef and in 1981, it became part of the World Heritage List. Since then, the first zoning plan for biodiversity protection and the regulation of activities within the Marine Park was developed. In 2015, a long-term plan was adopted for the coral reefs' sustainability through 2050.

For four decades, governments of Australia and the Province of Queensland, together with industries, scientists, community organizations and individuals, have invested significant resources for coral reef protection and management. These measures have been guided by the following principles: maintaining and reflecting the values of the World Heritage List in each action; underpinning decisions on scientific evidence; guaranteeing a net benefit for the ecosystem; adopting an associative management approach (Commonwealth of Australia, 2018).

Exclusive economic zone of Belgium

This is the most advanced MSP in Europe. The Belgian portion of the North Sea is one of the most heavily used sea passages in the world. Many activities converge on one another and could cause environmental impacts. Belgian authorities undertook an MSP process to create a balance between activities.

The MSP was started in 2003 with a master plan to establish specific areas for different coastal-marine productive activities. Given an international and European agenda focused on marine issues, UNESCO and IOC supported the MSP process. In 2011, the Belgian Marine Environmental Service led the MSP process and submitted their proposal in 2012, which was approved by Royal Decree later the same year.

This particular MSP process has also been informed by certain principles including: establishment of a spatial and temporary structure for the distribution of economic activities in the EEZ and adjacent coastal zones; balancing the presence of different activities undertaken within this geographical area; creating an "open" system with structural connections and cross-border relations with neighboring countries; identifying long-term strategic possibilities.

The MSP process was coordinated by the Federal Government with the participation of a wide range of actors, as provided in the Marine Environment Act. The first preliminary draft was submitted for a broad informal consultation.

An Advisory Committee later prepared a revised draft which was then reviewed by the Federal Council of Ministers. The environmental impact assessment report was also part of the process and was submitted for public consultation, including with Regional Governments, Coast Guards, the Federal Council for Sustainable Development and other relevant actors. Furthermore, the environmental impact assessment report was consulted with France, The Netherlands and Great Britain. A final report was reviewed by the Federal Council of Ministers and adopted through a Royal Decree (Belgian Federal Public Service Health, Food Chain Safety and Environment, 2015)

Source: Prepared by the author (2018)

Experiences in America

In the Americas there are several MSPs being implemented and a few others are in their conclusion phase. The most advanced plans are arguably related to the Galapagos Marine Reserve in Ecuador and an MSP in the state of Massachusetts in the United States of America.

According to UNESCO, plans in an implementation phase include: the Barbuda Marine Zoning (Antigua and Barbuda); the coastal areas, EEZ and territorial sea of Belize; four MSPs in Canada (Haida Gwaii, North Coast, Central Coast and North Vancouver Island) and the states of Rhode Island and Oregon in the United States (UNESCO, 2018a).

In the case of Mexico, three MSP processes have been concluded (for the Gulf of California, Gulf of Mexico and the Caribbean Sea, and the North Pacific) and one is under development (Central and Southern Pacific). The former MSPs have been formally approved and are ready for their implementation (Gulf of California since 2006; Gulf of Mexico and Caribbean Sea in 2012; and Northern Pacific in 2018).[5,6,7]

UNESCO also registers MSPs under way in Colombia, Costa Rica (Gulf of Nicoya) and Panama (Gulf of Montijo) (UNESCO, 2018a). Some countries are not on the UNESCO list but have developed other forms/instruments for marine and coastal planning based on ICZM, like Costa Rica (Marine Multiple Use Areas and Marine Areas for Responsible Fisheries), Dominican Republic (Ecosystem-Based Zoning in Samaná Bay) and Cuba (Protected Marine Areas and Zones Under an Integrated Coastal Management Regime) (Aldana and Hernández-Zanuy, 2016).

In the case of Cuba, of particular relevance is the implementation of integrated coastal management plans linked to the UNDP/GEF Sabana-Camagüey project (1993). The project promoted the creation of an Integrated Coastal Management Agency and introduced implementation experiences for ICZM in the Northern Province of Matanzas (Alcolado *et al.*, 2007).

Following is a brief analysis of an advanced case of MSP in Latin America (Galapagos Islands) and the current situation in Mexico, Chile and Peru.

Galapagos Islands

The interest in implementing MSP in the Galapagos Islands arises from the biological, environmental and economic importance of this territory, its recognition by UNESCO as the first World Heritage Site in 1979 and the existence of two natural protected areas: Galapagos National Park (1959) and the Marine Reserve (1998) established by the Organic Law for a Special Regime for the Conservation and Sustainable Development of Galapagos (DPNG, 2014).

The first planning efforts for the archipelago date back to 1974, when the National Planning Council in cooperation with FAO and UNESCO, developed a Master Plan for the Protection and Use of the Galapagos National Park. This has been followed by regular reviews of the Park's planning in 1984, 1996, 2005 and 2014. In the case of the Marine Reserve, the first plan was adopted in 1999 and its first review and adjustment was prepared in 2014 (DPNG, 2014). Three evaluations about the effectiveness of the National Park management have been undertaken in 1998, 2002 and 2012 and one for the Marine Reserve in 2011.

Critically important, the Management Plan for the Protected Areas of Galapagos for Good Living (2014) integrates, for the first time, land and marine territorial planning for the archipelago (DPNG, 2014). The Management Plan process was coordinated by the Directorate of the Galapagos National Park and took almost 15 months to be completed. Work was organized through a technical and core working group.

The first of these groups was formed by National Park and Marine Reserve personnel (mainly technicians and park rangers), as well as key social actors; the second group was composed of the Ministry of Environment (National Government), the Directorate of the Galapagos National Park, the government council, provincial governments, municipal autonomous decentralized governments (local level), parochial committees (community level) and the Agency of Biosafety and Quarantine for Galapagos (DPNG, 2014).

The Management Plan was inspired by the Decalogue of Good-Living and four principles therein: ecosystem management, conservation of ecosystems capacities to provide environmental services and their rational use, citizen participation and adaptive management (DPNG, 2014).

The Management Plan for the Protected Areas of Galapagos for Good Living integrates territorial planning for all the province of Galapagos – only 3 percent of the insular surface is not protected. The long-term vision is a common and shared forward-looking approach for the region, by controlling the impacts that human activities generate on the ecological integrity and resilience of ecosystems and encouraging endogenous socioeconomic development patterns, thus reducing dependency on continental Ecuador.

This approach has been defined as follows:

> The Province of Galapagos achieves good-living ["*buen vivir*"] for society by preserving its marine and insular ecosystems, and its biodiversity,

through a territorial model that integrates protected and populated areas. To achieve this, six objectives have been redrafted that need to be met through the implementation of the Management Plan.

(DPNG, 2014)[8]

Regional Marine Ecological Planning in the Mexican North Pacific

Just as marine ecological planning was taking place in the Gulf of Mexico in 2006 and Marine Spatial Planning was gaining traction internationally, Mexico approved a National Strategy for Ecological Planning of Seas and Coasts, to be applied in priority coastal municipalities.[9],[10] This is the most recent spatial planning process undertaken in Mexico and was finally approved in 2018.

The importance of this particular experience is that it was built as a step-by-step process, incorporated terrestrial dimensions and their more direct influence on coastal and marine zones, as was the case in the Gulf of Mexico and Caribbean Sea Marine Spatial Planning, and contributed with new methodologies, as proposed by Hernández de la Torre *et al.*, (2015).

Incorporating a terrestrial dimension in this MSP process has been significant, as it means compliance with existing ecosystem-based management commitments and watershed approaches and their application to Mexican marine zones and adjacent Federal zones.[11] This integration of terrestrial dimensions was reached through a coordination agreement between the Mexican Federation and federal agencies according to their competences. The Federal Government through the Secretariat of the Environment and Natural Resources coordinated the development of this agreement with 18 Federal entities and agencies as well as productive sectors and the conservation community.[12]

Ecological processes of the North Pacific Region of Mexico are of great importance. First because of their rich biodiversity and the fact they are transition zones between warm and tropical marine regions, where the Sub-Arctic Current, the California Current and North-Equatorial Current converge. Second, because of productive activities such as commercial fisheries, tourism and intensive agriculture, which impact ecosystems.[13]

The main problems that need to be addressed by the MSP process are overfishing and illegal fishing, modification of wetlands, the destruction and alteration of dunes and coastal lagoons, the destruction and alteration of mangroves, obstruction of access to beaches, pollution from municipal and industrial waste water, extraction of stone materials, erosion, mining in marine zones, salinization of aquifers and occupation of the coastal strip.[14]

The situation of Marine Spatial Planning in Chile

The academic sector in Chile has played a critical role in marine conservation, planning and management. Academia was first to propose the need for marine biodiversity protection and the adoption of an holistic approach based on scientific knowledge, to establish a national network of protected areas. Various coastal and marine conservation programs focus on Marine Spatial

Planning and establishment of marine protected areas (Advanced Conservation Strategies, 2011).

Although the enactment of the General Law on Fisheries and Aquaculture in 1991, created the first legal framework for the conservation of marine spaces and ecosystems, including through marine parks and reserves and a System for Management and Exploitation Areas for Benthic Resources (Araos *et al.*, 2017), it was the National Policy for the Use of the Coastal Border in 1994, which opened a new phase for coastal and marine planning. It created a formal decision-making structure with the participation of multiple actors, both public and private (IMARPE *et al.*, 2008).

In 2004, the first Multiple Use Marine Protected Areas were created with the support of the United Nations Development Program (UNDP) (Sierralta *et al.*, 2011). This enabled the sustainable use of marine resources based on a shared administration and management model, which promotes the participation of public and private sectors in their elaboration, implementation and development (de Andrade *et al.*, 2010).

In 2008, a new law created the Coastal and Marine Spaces of Native Peoples, which recognizes indigenous peoples as ancestral inhabitants and users of the coastal border. By 2015, three Coastal and Marine Spaces had been approved and created, eight more were under review and 35 applications had been presented to the competent authority (Gissi *et al.*, 2017).

Starting in 2010, local initiatives for marine conservation and coastal management at the municipal level were undertaken with a view to generate instruments for the protection and management of the coastal natural heritage of municipalities, through the coordination of different social actors such as municipal governments, artisanal fishers, academics and locally based NGOs (Araos *et al.*, 2017).

Diverse MSP processes have been undertaken in recent years. The UNESCO "Man and the Biosphere" Programme, with the support of the Government of Flanders, Belgium undertook a process in Chiloé in southern Chile (UNESCO, 2017). The process strongly integrated scientific knowledge and strengthen general outcomes (Tognelli *et al.*, 2009; Outeiro *et al.*, 2015). Another process included the Magallanes region and Antarctica, as a response to its establishment as a priority area for tourism development (Nahuelhual *et al.*, 2017).

The mapping of ecosystem services – absent in the early part of this decade – has been an important input to integrate spatial planning with marine planning and management (Böhnke-Henrichs *et al.*, 2013). Linked to spatial planning processes at the regional level, some efforts have also been made by academia and non-governmental organizations to support planning at the community level, for example with the Quemchi Commune, Chiloé (Montenegro, 2010).

The situation of Marine Spatial Planning in Peru

In the case of Peru, concerns for marine conservation are rather recent. Historically conservation priorities have focused on the Andes and Amazon and capacities, resources and activities in marine conservation activities and promotion of sustainable fisheries have been limited to efforts by a few NGOs. In recent years, more national and international NGOs have become actively engaged in marine and fisheries conservation. Likewise, academia and the private sector have shown increased interest in conservation of marine and coastal areas (Advanced Conservation Strategies, 2014).

A series of activities have also been undertaken over the past decades to protect marine and coastal spaces. Examples include, the Action Plan for the Protection of the Marine Environment and Coastal Areas of the South East Pacific (1981), which led to the Environmental Planning of the Pisco–Paracas Area, concluded in 1998, or the pilot project Recovery of the Ferrol Bay and Adjacent Zones in 2005 (IMARPE *et al.*, 2008). With regards to the integration of coastal and marine zone management into broader strategies, the Binational Program for the Integrated Management of the Humboldt Current Large Marine Ecosystems created to promote integrated planning and management of the Callao coastal marine zone is a milestone in Peru (Cabrera *et al.*, 2005).

Another example at the international/regional level is UNESCO and IOC project Southeast Pacific Data and Information Network in Support to Integrated Coastal Area Management sponsored by the Permanent Commission for the South Pacific (IMARPE *et al.*, 2008). The Biosphere Reserves as a Tool for Coastal and Island Management in the South-East Pacific project coordinated by the UNESCO's Man and the Biosphere Program is also a key standout.

As a result of Peru's participation in both these initiatives are inclusion of a coastal zone in the Northwest Biosphere Reserve, the elaboration of a coastal and marine atlas to support decision-making and development of a Sustainable Blue Growth Strategy for Fisheries and Aquaculture for Sechura Bay, Piura, in northern Peru (UNESCO, 2018b) where a regional program was implemented in 2006 (Gobierno Regional de Piura, 2006) and led to the pilot case in 2014 (Gobierno Regional de Piura, 2014).

Although MSP in Peru is an ongoing learning process, it is an important component in the national agenda for marine governance. It has been consolidating and this is best reflected in the approval of Guidelines for Integrated Coastal Zone Management which seeks to inform marine and coastal management national policy through a planning process (Barragán and Lazo, 2018).

The first step for effectiveness in MSP involves updating the map of all coastline uses, pressures and driving forces in order to allow the identification of friction points and potential conflict sites. Research efforts should be focused on community participation and social learning through a development process (McKinley *et al.*, 2018).

Conclusions

Marine Spatial Planning has become a universally accepted tool and instrument. However, its interpretation, development and tailoring to national needs and impacts varies substantially across regions.

Two key elements in MSP include an integrated approach to developing coastal and marine spaces, multi-purpose objectives, multi-sectorial participation, long-term approaches and coordination among different levels of government.

Marine Spatial Planning is seen as a positive tool to enhance conservation of marine and coastal biological diversity, design of natural protected areas and sustainable management of resources in marine and coastal zones. Many successful examples evidence positive environmental, economic and social perspective impacts.

Marine Spatial Planning also plays a key role in achieving Objective 14 of the Sustainable Development Goals (SDG). This tool assists maintenance of biological diversity, conservation of ecosystem services, favors the balance among marine and coastal zone activities, encourages interinstitutional coordination, generates synergies among different processes and sectors, improves the quality of life for populations and promotes local appropriation for planning.

In the case of developing countries, challenges for MSP include appropriate financing, pressure from economic sectors and fragile democracies which affect both planning and implementation, including monitoring and improving processes.

The case of the Galapagos Islands is unique. It has passed three evaluation processes and has adapted accordingly to new circumstances and progressed to a third level in the MSP SP cycle proposed by UNESCO.

Further limitations include limited or suboptimal coordination between different orders of government, in addition to only partially sharing the commitment to the forward-looking approach.

Out of the ten steps established in the UNESCO/IOC Guide for Marine Spatial Planning, four critical points are worth highlighting. First, a leader and champion for the process is required. This could be an institution with widespread credibility and acceptance among participant actors, and moral authority to conduct the process. Second, obtaining funding, that is becoming increasingly limited, demands exploration of funding schemes with public, private and international participation. Third, constructing by consensus a long-term forward-looking approach is crucial as well as ensuring joint responsibility among participants in order to implement all the necessary actions. Finally, organizing effective participation of actors involved that leads to the social appropriation of the instrument, is the only way to guarantee adequate implementation of a spatial management plan.

Final reflections

Although MSP has been shown to be important to promote sustainable development, conserve biological diversity and maintain the services provided by ecosystems, there are some relevant aspects that need to be improved. These conditions can vary widely according to circumstances, and scaling-up and replicability should, if possible, respond to these changing conditions.

One key aspect to consider is the issue of participation of actors and communities that live in and use the coastal and marine space subject to planning. It is not the same to organize a symbolic participation process, where actors and communities are simply informed or consulted on the spatial management plan proposal, rather than to create effective and informed participation spaces and mechanisms that promote association among actors, the delegation of functions, joint responsibility and appropriation of the process and products generated by those involved. This will allow real participatory governance associated with MSP and true participation.

A second point that appears as a bottleneck is the construction of a long-term forward-looking approach. It is evident that in developed countries with consolidated democracies, strategic planning and consensus are, to a considerable extent, aligned with customs and traditions constructed over time among different political groups, economic actors, civil organizations and citizens. The situation in developing countries is more fragile. "Long-term" ends with the administrative period of the president in office. This huge difference of what can be understood as "long-term" does not allow planning processes to comply with the continuous adaptive cycle proposed by UNESCO/IOC, where a new revised plan is elaborated based on evaluation results and the research applied.

The practical consequence of this is that a long-term administrative approach is really a short- or medium-term approach where economic and group interests predominate over the interests of communities, less powerful economic actors. This has an effect on sustainable development possibilities. A way to overcome this is to include short-, medium- and long-term goals in the MSP process through consensus among participating actors, in order to reach a strategic approach at the local level that can survive presidential terms.

A third point to mention is directly related to the impossibility of complying with the continuous adaptive cycle proposed by UNESCO/IOC. Low probability of passing to a second level of planning through the development of a new revised plan based on evaluation results and the research applied, where a long-term administrative approach prevails, is a continued limitation.

In many developing countries the evaluation is not conducted because no resources are available or because the administrative period of the government in office has ended, or if it is undertaken, the government does not use the evaluation results due to limitations in mediation and/or the scope of the evaluation, or from the lack of capacity to insert results obtained into the policy process and organizational dynamics (Sosa, 2011).

This situation causes protracted planning cycles, where it is practically impossible to go past the first level. An exceptional case is the Galapagos Archipelago, where international interests have played a fundamental role, in addition to the sensitivity all presidents of Ecuador that served from 1974 to date who have had a true interest and commitment to supporting actions and measures for this geographical space.

One alternative is to establish an association among producer organizations, investors, universities, research centers and government authorities, to finance and conduct necessary evaluations with the commitment of using the results in a next level of the adaptive cycle of the MSP process.

Finally, it is important to reflect on decision-making principles specifically established for the Great Barrier Reef Australia planning process. This implies recognition of universal values of a world heritage site; supporting decisions on the best knowledge available, including the traditional knowledge of original peoples; consideration of risks associated to climate change; acknowledging net benefits for ecosystems, underpinned by sustainable development principles, including the precautionary principle; avoiding environmental impacts; mitigating residual impacts and restoring the resilience and health of ecosystems; and lastly, adopting an associative approach among actors for management. This demands transparent governance processes which encourage a wide range of opportunities for sustainable development, including traditional use, promoting cooperation, the empowerment of actors, delegating authority to local and community governments, as well as adopting innovations in the management of the space and its resources.

If we have the capacity to adopt the four principles for decision-making in Marine Spatial Planning processes, we would be enhancing potential benefits of results, as well as constructing strong community-based support through the construction of governance structures which also respond to MSP objectives and the continuous adaptive cycle proposed by UNESCO/IOC.

Acknowledgments

I would like to thank the collaboration of María Fernanda Onofre Villalva and Mariana Torres García, students of the Bachelor Program on Environmental Sciences of the National School of Superior Studies of the Autonomous National University of Morelia, México.

Notes

1 Environmental degradation is the deterioration of the environment through depletion of resources such as air, water and soil; the destruction of ecosystems and the extinction of wildlife. It is defined as any change or disturbance to the environment perceived to be deleterious or undesirable, or as the set of processes that deteriorates or prevents the use of a determined resource. See, Zurrita, A.A., Baddi, M.H., Guillen, A., Lugo Serrato, O., Aguilar Garnica, J.J. 2015. Factores Causantes de Degradación Ambiental. Daena: *International Journal of Good Conscience* 10(3): 1–9.

2 It is estimated that two-thirds of the planets' megalopolises are found on coastal zones. See, UNEP, FAO, IMO, UNDP, IUCN, World Fish Center, GRID-Arendal. 2012. *Green Economy in a Blue World*. United Nations Environment Programme. Nairobi.

3 One of the most emblematic examples is that of Chesapeake Bay in Virginia U.S.A. After using marine and coastal resources for 300 years, fisheries were on a dramatic decline over the last decades of the twentieth century. With the incorporation of an ecosystem-based management approach and the participation of

academia and governmental agencies, as well as the users of resources and local people, fisheries stocks have been recuperated through ecosystem-based fisheries planning. See, Chesapeake Bay Fisheries Ecosystem Advisory Panel (National Oceanic and Atmospheric Administration Chesapeake Bay Office). 2006. *Fisheries Ecosystem Planning for Chesapeake Bay*. American Fisheries Society, Bethesda. Another example is the Lesser Sunda Archipelago in Indonesia. It is one of the richest regions of biological and cultural diversity, threatened by global climate change and affected by intensive destructive fishing, marine contamination and coastal development. Since 2008, an international organization with German funding has provided support to the Indonesian Government to implement ecosystem-based management and incorporate sustainable management of human activities criteria and measures. Three strategies were developed in parallel for marine and coastal planning: the multi-objective MSP, a network of marine protected areas and marine conservation agreements. See: Perdanahardja, G. and Lionata, H. (2017), *Nine Years in Lesser Sunda. Indonesia.* The Nature Conservancy and Indonesia Coasts and Oceans Program, Kebayoran Baru.

4 The Guide divides the process into ten steps: 1) identify the needs and establish the authority in charge of directing the process, 2) obtain funding, 3) organize the process by means of pre-planning, 4) organize the participation of actors, 5) definition and analysis of existing conditions, 6) definition and analysis of future conditions, 7) elaboration and approval of the spatial management plan, 8) implement and establish the spatial management plan, 9) monitor and evaluate the performance, and 10) adaptation of the marine spatial management process. See: Douvere, F. and Ehler, C.N. 2011. The Importance of Monitoring and Evaluation in Adaptive Maritime Spatial Planning. *Journal of Coastal Conservation* 15(2): 305–311.

5 SEMARNAT (2006a), Accord that creates the Program for Marine Planning for the Gulf of Mexico and Ecologic Management for the Gulf of California. *Diario Oficial de la Federación* 15 December 2006.

6 SEMARNAT (2012), Accord that creates the marine section of the Program for Regional Marine Planning for the Gulf of Mexico and the Caribbean Sea and the regional aspect of the Program is revealed. *Diario Oficial de la Federación* 24 November 2012.

7 SEMARNAT (2018), ACUERDO por el que se da a conocer el Programa de Ordenamiento Ecológico Marino y Regional del Pacífico Norte. *Diario Oficial de la Federación* 9 August 2018.

8 The Management Plan has as its basic objectives: OB1. Manage ecosystem conservation; OB2. Incorporate protected areas policies into the ordering plan; OB3. Improve management capacity of the Directorate of the Galapagos National Park; OB4. Revitalize participatory social processes; OB5. Increase scientific knowledge integrating it in decision-making; and OB6. Promote national and international cooperation.

9 SEMARNAT. (2007), *Estrategia Nacional para el Ordenamiento Ecológico del Territorio en Mares y Costas. Subsecretaría de Planeación y Política Ambiental.* Dirección General de Política Ambiental e Integración Regional y Sectorial.

10 *Op. cit.* See note 5.

11 SEMARNAT. (2014), Reglamento de la Ley General del Equilibrio Ecológico y la Protección al Ambiente en Materia de Ordenamiento Ecológico. Última reforma. *Diario Oficial de la Federación* 31 October 2014.

12 *Op. cit.* See note 7.

13 In accordance with current legislation, the Ecological Planning Processes in Mexico must comply with four basic provisions to elaborate the program proposal: 1) methodological rigor when obtaining information, analyzing and generating results; 2) transparency when obtaining information and generating results; 3) be systematic in the verification of results presented; and 4) include the participation of main sectors of society that come into play with the distribution of activities and land use (SEMARNAT, 2006b).
14 *Op. cit.* See note 7.

References

Advanced Conservation Strategies. (2011), *A Coastal-Marine Assessment of Chile.* A Report Prepared for the The David and Lucile Packard Foundation.

Advanced Conservation Strategies. (2014), *A Marine Conservation Assessment in Peru.* A Report Prepared for The David and Lucile Packard Foundation and Foundation Ensemble.

Alcolado, P.M., García, E.E. and Arellano-Acosta, M. (Eds). (2007), *Ecosistema Sabana-Camagüey. Estado Actual, Avances y Desafíos en la Protección y Uso Sostenible de la Biodiversidad.* Editorial Academia. La Habana.

Aldana, O. and Hernández-Zanuy, A.C. (2016), La Planificación Espacial Marina: Marco Operativo para Conservar la Diversidad Biológica Marina y Promover el Uso Sostenible del Potencial Económico de los Recursos Marinos en el Caribe. In: Hernández Zanuy, A.C. (Ed.). *Adaptación Basada en Ecosistemas: Alternativas para la Gestión Sostenible de los Recursos Marinos y Costeros del Caribe.* Instituto de Oceanología de Cuba. La Habana, pp. 109–121.

Ansong, J., Gissi, E. and Calado, H. (2017), An Approach to Ecosystem–Based Management in Maritime Spatial Planning Process. *Ocean & Coastal Management,* 141: 65–81.

Araos, F., Godoy, C., Andrade, R., Ther, F., Gelcich, S. and Salas, C. (2017), Conservación Marina y Costera en Chile: trayectorias institucionales, innovaciones locales y recomendaciones para el futuro. In: Ferreira, L. Schmidt, L., Pardo, M., Calvimontes, J. and Viglio, E. (Eds). *Cima de Tensão. Ação humana, biodiversidade e mudanças climpaticas.* UNICAMP. Campinas, pp. 529–544.

Barragán, J.M. and Lazo, O. (2018), Policy progress on ICZM in Peru. *Ocean and Coastal Management,* 157: 203–216.

Baskett, M.L., Gaines, S.D. and Nisbet, R.M. (2009), Symbiont Diversity May Help Coral Reefs Survive Moderate Climate Change. *Ecological Application,* 19(1): 3–17.

Belgian Federal Public Service Health, Food Chain Safety and Environment. (2015), *Marine Spatial Plan for the Belgian Part of the North Sea.* Federal Public Service Public Health, Food Chain Safety and Environment. Brussels.

Belgian State. (2012), *Socio-Economic Analysis of the Use of the Belgian Marine Waters and the Costs Associated with the Damage to the Marine Environment.* Marine Strategy Framework Directive – Art 8, lid 1c. Federal Public Service Public Health, Food Chain Safety and Environment, Brussels, Belgium.

Böhnke-Henrichs, A., Baulcomb, C., Koss, R. and Salman Hussain, S. (2013), Typology and Indicators of Ecosystem Services for Marine Spatial Planning and Management. *Journal of Environmental Management,* 130: 135–145.

Botero, C.M., Fanning, L.M., Milanes, C. and Planas, J.A. (2016), An Indicator Framework for Assessing Progress in Land and Marine Planning in Colombia and Cuba. *Ecological Indicators*, 64: 181–193.

Buhl-Mortensen, L., Galparsoro, I., Vega Fernández, T., Johnson, K., D'Anna, G., Badalamenti, F., ... Doncheva, V. (2017), Maritime Ecosystem-Based Management in Practice: Lessons Learned from the Application of a Generic Spatial Planning Framework in Europe. *Marine Policy* 75: 174–186.

Cabrera, C., Maldonado, M., Arévalo, W., Pacheco, R., Giraldo, A. and Quispe, J. (2005), Planificación y Gestión Integrada de la Zona Marina Costera del Callao. *Revista del Instituto de Investigación FIGMMG* 8(16): 38–43.

Calderón-Aguilera, L.E., Reyes-Bonilla, H., Norzagaray-López, C.O. and López-Pérez, R.A. (2017), Los Arrecifes Coralinos de México: Servicios Ambientales y Secuestro de Carbono. *Elem. para Políticas Públicas* 1(1): 53–62.

Carballo, J.L., Bautista-Guerrero, E., Nava, H. and Cruz-Barraza, J.A. (2010), Cambio Climático y Ecosistemas Costeros. Bases Fundamentales para la Conservación de los Arrecifes de Coral del Pacífico Este. In: Hernández–Zanuy, A., Alcolado, P.M. (Eds.), *La Biodiversidad en Ecosistemas Marinos y Costeros del Litoral de Iberoamérica y el Cambio Climático: I. Memorias del Primer Taller de la RED CYTED-Biodivamr.* CYTED. La Habana. pp. 183–193.

Chesapeake Bay Fisheries Ecosystem Advisory Panel (National Oceanic and Atmospheric Administration Chesapeake Bay Office). (2006), *Fisheries Ecosystem Planning for Chesapeake Bay*. American Fisheries Society, Bethesda.

Commonwealth of Australia. (2018), *Reef 2050 Long-Term Sustainability Plan – July 2018*. Australian Government, Queensland Government.

De Andrade, R., Cabezas, A., Cornejo, S., Godoy, C., Moreno, M. and Villablanca, R. (2010), Guía de modelos de administración y gestión participativa de Áreas Marinas y costeras Protegidas de Múltiples Usos (AMCP-MU). PNUD. Santiago.

Domínguez-Tejo, E., Metternicht, G., Johnston, E. and Hedge, L. (2016), Marine Spatial Planning Advancing the Ecosystem-Based Approach to Coastal Zone Management: A Review. *Marine Policy* 72: 115–130.

Donner, S.D., Heron, S. and Skirving, W.J. (2008), Future Scenarios: A Review of Modelling Efforts to Predict the Future of Coral Reefs in an Era of Climate Change. In: van Oppen, M.J.H. and Lough, J.M. 2009. *Coral Bleaching: Patterns, Processes, Causes and Consequences*. 159–173. Springer-Verlag. Berlin, Heidelberg.

Douvere, F. and Ehler, C.N. (2011), The Importance of Monitoring and Evaluation in Adaptive Maritime Spatial Planning. *Journal of Coastal Conservation* 15(2): 305–311.

Dirección del Parque Nacional Galápagos (DPNG). (2014), *Plan de Manejo de las Áreas Protegidas de Galápagos para el Buen Vivir*. Ministerio del Ambiente, DPNG y WWF. Puerto Ayora.

Ehler, C. and Douvere, F. (2009), *Marine Spatial Planning: A Step-By-Step Approach Toward Ecosystem-Based Management*. Intergovernmental Oceanographic Commission and Man and the Biosphere Programme. IOC Manuals and Guides No. 53, ICAM Dossier No. 6. UNESCO. Paris.

Gissi, N., Ibacache, D., Pardo, B. and Ñancucheo, M.C. (2017), El estado chileno, los lafkenche y la Ley 20,249: ¿Indigenismo o Política de Reconocimiento? *Revista Austral de Ciencias Sociales* 32: 5–21.

Gobierno Regional de Piura. (2006), *Programa Regional de Manejo Integrado de Recursos de la Zona Marino Costera de Piura*. Gobierno Regional de Piura. Piura.

Gobierno Regional de Piura. (2014), *Plan de Manejo Integrado de los Recursos de la Zona Marino Costera de Sechura*. Gobierno Regional de Piura, Pro Gobernabilidad. Piura.

Heinrich Böll Stiftung (HBS). (2017), *Atlas de los Océanos 2017*. HBS. Ciudad de México.

Hernández de la Torre, B., Aguirre Gómez, R., Gaxiola-Castro, G., Álvarez Borrego, S., Gallegos-García, A., Rosete Vergés, F. and Bocco Verdinelli, G. (2015), Ordenamiento Ecológico Marino en el Pacífico Norte Mexicano: Propuesta Metodológica. *Hidrobiológica* 25(2): 151–163.

Iglesias-Campos, A., Ehler, C. and Barbière, J. (2014), *Enfoque Global de la Gestión Costera y la Planificación Espacial Marina*. Reunión sobre el Futuro de SPINCAM. UNESCO/COI. Guayaquil.

Instituto del Mar de Perú (IMARPE), UNESCO, COI, Gobierno de Flandes and Comisión Permanente del Pacífico Sur (CPPS). (2008), *Informe del Taller sobre Experiencias en el Desarrollo de Indicadores de Gestión en Manejo Costero Integrado en los Países del Pacífico Sudeste*. IMARPE. Lima.

Intergovernmental Panel on Climate Change (IPCC) (2018), *Global Warming of 1.5°C. An IPCC Special Report on the Impacts of Global Warming of 1.5°C above Pre-Industrial Levels and Related Global Greenhouse Gas Emission Pathways, in the Context of Strengthening the Global Response to the Threat of Climate Change, Sustainable Development, and Efforts to Eradicate Poverty*. Summary for Policymakers. WMO, UNEP. Switzerland.

Kritzer, J.P., Delucia, M.B., Greene, E., Shumway, C., Topolski, M.F., Thomas-Blate, J., Chiarella, L.A., Davy, K.B. and Smith, K. (2016), The Importance of Benthic Habitats for Coastal Fisheries. *Bioscience* 66(4), 274–284.

Long, R.D., Charles, A., Stephenson, R.L. (2015), Key Principles of Marine Ecosystem-Based Management. *Marine Policy* 57: 53–60.

McKinley, E., Aller-Rojas, O., Hattman, C., Germond-Duret, C., Vicuña San Martín, I., Hopkins, C. R., Aponte, H. and Potts, T. (2018), Charting the Course for a Blue Economy in Peru: A Research Agenda. *Environment, Development and Sustainability* https://doi.org/10.1007/s10668-018-0133-z.

Montenegro, F.A. (2010), *Análisis Territorial Integrado y Propuesta de Ordenamiento Territorial de la Zona Costera de la Comuna de Quemchi. Una Integración de la Planificación Ecológica y Participativa con Miras al Desarrollo Local*. Thesis. Universidad de Chile. Santiago.

Muñoz, N.P. and Le Bail, M. (2017), Latin American and Caribbean Regional Perspective on Ecosystem Based Management (EBM) of Large Marine Ecosystems Goods and Services. *Environmental Development* 22: 9–17.

Nahuelhual, L., Vergara, X., Kusch, A., Campos, G. and Droguett, D. (2017), Mapping Ecosystem Services for Marine Spatial Planning: Recreation Opportunities in Sub-Antarctic Chile. *Marine Policy* 81: 211–218.

Nava, J.C., Arenas, P., Cardoso, F. (2018), Integrated Coastal Management in Campeche, Mexico: A Review after the Mexican Marine and Coastal National Policy. *Ocean and Coastal Management* 154: 34–45.

Organización de las Naciones Unidas (ONU). (2012), *Documento final de la Conferencia sobre el Desarrollo Sostenible*. Río de Janeiro, Brasil.

Organización de las Naciones Unidas (ONU). (2016), *Agenda 2030 y los Objetivos del Desarrollo Sostenible. Una oportunidad para América Latina y el Caribe.* CEPAL. Santiago de Chile.

Outeiro, L., Häussermann, V., Viddi, F., Hucke-Gaete, R., Försterra, G., Oyarzo, H., Kosiel, K. and Villasante, S. (2015), Using Ecosystem Services Mapping for Marine Spatial Planning in Southern Chile under Scenario Assessment. *Ecosystem Services* 16: 341–353.

Pauly, D., Christensen, V., Guénette, S., Pitcher, T.P., Sumaila, U.R., Walters, C., Watson, R. and Zeller, D. (2002), Towards Sustainability in World Fisheries. *Nature* 418: 689–695.

Perdanahardja, G. and Lionata, H. (2017), *Nine Years in Lesser Sunda. Indonesia.* The Nature Conservancy and Indonesia Coasts and Oceans Program, Kebayoran Baru.

Pomeroy, R.S., Baldwin, K. and McConney, P. (2014.), Marine Spatial Planning in Asia and the Caribbean: Application and Implications for Fisheries and Marine Resource Management. *Desenvolvimento e Meio Ambiente* 32: 151–164.

Pondella, D., Schiff, K., Schaffner, R., Zellmer, A. and Coates, J. (2016), *Southern California Bight 2013 Regional Monitoring Program: Volume II. Rocky Reefs.* SCCWRP Technical Report 932.

Resolución Ministerial No. 189–2015-MINAM. (2015), *Aprobación de los "Lineamientos para el Manejo Integrado de las Zonas Marinos Costeras".* Ministerio del Ambiente de Perú. Lima.

Rodriguez, N.J.I. (2017), A Comparative Analysis of Holistic Marine Management Regimes and Ecosystem Approach in Marine Spatial Planning in Developed Countries. *Ocean & Coastal Management* 137: 185–197.

Rosete, F.A. and Fuentes, J.J. (Eds). (2018), *Aportes de la Planeación Territorial en Hispanoamérica. Estudios de caso Desde Diferentes Perspectivas.* ENES-UNAM. Morelia, Mexico.

SEMARNAT (2006a), Accord that creates the Program for Marine Planning for the Gulf of Mexico and Ecologic Management for the Gulf of California. *Diario Oficial de la Federación* 15 December.

SEMARNAT. (2006b), *Manual para el Proceso de Ordenamiento Ecológico.* Secretaría de Medio Ambiente y Recursos Naturales, México.

SEMARNAT. (2007), *Estrategia Nacional para el Ordenamiento Ecológico del Territorio en Mares y Costas. Subsecretaría de Planeación y Política Ambiental.* Dirección General de Política Ambiental e Integración Regional y Sectorial, Mexico, DF.

SEMARNAT. (2012), Accord that Creates the Marine Section of the Program for Regional Marine Planning for the Gulf of Mexico and the Caribbean Sea and the Regional Aspect of the Program is Revealed. *Diario Oficial de la Federación* 24 November 2012.

SEMARNAT. (2014), Reglamento de la Ley General del Equilibrio Ecológico y la Protección al Ambiente en Materia de Ordenamiento Ecológico. Última reforma. *Diario Oficial de la Federación* 31 October 2014.

SEMARNAT (2018), ACUERDO por el que se da a conocer el Programa de Ordenamiento Ecológico Marino y Regional del Pacífico Norte. *Diario Oficial de la Federación* 9 August 2018.

Sierralta, L., Serrano, R., Rovira, J. and Cortés, C. (Eds). (2011), *Las Areas Protegidas de Chile.* Ministerio del Medio Ambiente. Santiago.

Sosa, J. (2011), Bases conceptuales para el Abordaje de la Política Federal de Evaluación en México: Una Propuesta Analítica. In: Ramos, J.M., Sosa, J. and Acosta, F. (Eds), *La evaluación de políticas públicas en México*. INAP-COLEF. Tijuana. pp. 101–120.

Spalding, M.D., Ravilious, C. and Green, E.P. (2001), *World Atlas of Coral Reefs*. Prepared at the United Nations Environment Programme – World Conservation Monitoring Centre. University of California Press. Berkeley.

Teneva, L., Karnauskas, M., Logan, C.A., Bianucci, L., Currie, J.C. and Kleypas, J.A. (2011), Predicting Coral Bleaching Hotspots: The Role of Regional Variability in Thermal Stress and Potential Adaptation Rates. *Coral Reefs* 31(1): 1–12.

Tognelli, M.F., Fernández, M. and Marquet, P.A. (2009), Assessing the Performance of the Existing and Proposed Network of Marine Protected Areas to Conserve Marine Biodiversity in Chile. *Biological Conservation* 142: 3147–3153.

UNEP, FAO, IMO, UNDP, IUCN, World Fish Center, GRID-Arendal. (2012), *Green Economy in a Blue World*. United Nations Environment Programme. Nairobi.

UNEP and GRID-Arendal. (2016), *Marine Litter Vital Graphics*. United Nations Environment Programme and GRID-Arendal. Nairobi and Arendal.

UNESCO. (2017), *Biosphere Reserves as a Tool for Coastal and Island Management in the South-East Pacific Region (BRESEP)*. Accessed 24 December 2018. www.unesco.org/new/en/natural-sciences/environment/ecological-sciences/specific-ecosystems/island-and-coastal-areas/bresep/.

UNESCO. (2018a), *Status of MSP*. Accessed 24 December 2018. http://msp.ioc-unesco.org/world-applications/status_of_msp/.

UNESCO. (2018b), *Resultados de la Colaboración entre los Proyectos BRESEP y SPINCAM en la Costa del Pacífico Sur de América Latina*. UNESCO. Paris.

Veron, J.E.N., Stafford-Smith, M.G., Turak, E. and DeVantier, L.M. (2018), *Corals of the World*. Accessed 24 December 2018, Version 0.01 (Beta). http://coralsoftheworld.org/v0.01 (Beta). Current version: http://coralsoftheworld.org.

World Wildlife Fund (WWF). 2018. *Informe Planeta Vivo-2018*: Apuntando más alto-Resumen. Grooten, M. and Almond, R.E.A. (Eds). WWF. Gland.

Zurrita, A.A., Baddi, M.H., Guillen, A., Lugo Serrato, O. and Aguilar Garnica, J.J. (2015), Factores Causantes de Degradación Ambiental. *Daena: International Journal of Good Conscience* 10(3): 1–9.

8 From customary law to the implementation of safeguard measures

The case of "Marine and Coastal Areas for Indigenous Peoples" in Chile

Luciano Hiriart-Bertrand, José Manuel Troncoso, Carlos I. Vargas and Alejandro Correa

Introduction

Coastal and marine environments are complex and unique spaces, which involve fragile and dynamic ecosystems where different types of uses commonly interact (Andrade, 2000; Barragan, 2003). The presence of abundant natural resources explains the importance and complexity of the coastal boarder. Multiple actors and interests from various sectors converge on the coast and shoreline, producing constant tension with and among coastal communities. For centuries, different communities and human settlements have made use of these coastal and marine spaces and existing resources. Areas located on the coastal boarder are of significant importance for the development of different cultures associated with the ocean and seas (Rugebret, 2015). The rapid increase and diversification of economic activities in coastal areas, has contributed to the rise of multiple conflicts involving existing property rights and uses of lands. Increases in the population density has meant more pressure exerted on coastal spaces, negatively impacting these areas and therefore affecting the livelihoods and means of subsistence of coastal communities (Gadgil *et al.*, 1993). As a result, unplanned development has been linked to the exploitation of resources and deterioration of ecosystems. In parallel with a historic pattern of violation of customary rights of indigenous peoples and local communities in Latin America in general, this situation has encouraged continued territorial vindication processes in Chile (Miller, 2012).[1]

At the same time, socio-environmental conflicts in Chile have been driven largely by weak state institutions and the absence of land-use and planning instruments.[2] This scenario worsens in rural coastal areas, where legislation is often indicative and sectorized. As in many countries, there is an administrative-based approach which prevails over coastal territorial planning and which does not respond appropriately to the multiple dimensions affecting coastal areas and the range of interests of actors that inhabit them (Schlotfeldt, 2000). As a consequence of a fragmented multi-sectorial bureaucracy that manages the coastal borders and the diversity of activities, actors, interests and regulations pertaining to development and, at the same time, affects the recognition of customary uses and rights of local communities, emerges the need to introduce a land-use planning instrument with a broader and more inclusive perspective of coastal and marine areas. In this regard, this chapter analyses the role of Marine and Coastal Areas for Indigenous Peoples (MCAIP) (Espacios Costero Marino de Pueblos Originarios), as a new system of sociocultural recognition which is a key pillar in safeguarding customary uses by coastal communities or indigenous peoples. The chapter explores a series of conflicts over time and how they have been addressed in order to advance the recognition of customary rights as part of a collective and historic vindication process.

Institutional and legal framework for the Chilean coastal border

The main instrument that currently governs the management of the coastal border in Chile is the National Policy for Use of the Coastal Border (Política Nacional de Uso del Borde Costero, PNUBC), introduced in 1994 and incorporated under the competence of the Ministry of Defense, in charge of the administration and exclusionary control of such spaces. The National Policy for Use of the Coastal Border defines its main objectives as: i) to promote the development of resources and wealth in different sectors, and ii) to protect and conserve the marine, land and air environment, in line with development needs and the other existing policies.

The PNUBC focuses on productive management and development, but recognizes ecological, social and cultural objectives and emphasizes ecological sustainability as a policy pillar. However, the PNUBC also evidences notorious internal contradictions when it addresses preferred uses of the coastal border, providing explicit preference to the following uses as examples of "better use": a) ports and other port facilities; b) construction and repair industries for vessels; c) regularization of human settlements and existing artisanal fishing inlets; d) public use areas for social recreation and entertaining purposes; and e) industrial, economic and development activities such as tourism, fishing, aquaculture, fisheries and mining.

The PNUBC is implemented through Coastal Border Zoning (Zonificación del Borde Costero), in charge of Regional Commissions for the Use of Coastal Borders (Comisiones Regionales de Uso del Borde Costero, CRUBC). Despite its potential as a tool for the harmonious development of the coastal border through the prevention of conflicts and a balanced distribution of environmental burdens, its effectiveness has been limited, mostly used as a guiding reference, in contrast with similar planning instruments (Serani, 2013). Thus, zoning has not been successful nor fully implemented. A critical aspect in the management of coastal borders is that even though the CRUBC play a critical role in their management, definition of uses and conflict prevention, their functions are merely consultative and their guidance has no binding effects on the agencies involved and users interested in this particular space.

Access rights of coastal indigenous peoples

The violation of rights reflected often in restrictions to territorial control and resulting management of resources by indigenous peoples dates back to the time of colonization (Miller, 2012). The relationship between the state of Chile and indigenous peoples has not been exempt from conflicts, including present times (Agostini *et al.*, 2010; Duquesnoy, 2012). In 1989, the Agreement Nueva Imperial was signed between the Christian Democratic presidential candidate Patricio Aylwin and Mapuche leaders, promising a new relationship between the state and indigenous peoples (Boccara and Seguel-Boccara, 1999). When democracy was regained in 1990, indigenous peoples – particularly the Mapuche – voiced several demands which focused on: the constitutional recognition of ethnic and cultural diversity; participation of their representatives in conducting the state's indigenous policies; the creation of parliamentary seats exclusively for indigenous representatives; the legal protection of their land and water; the return of public or private land taken over by the state; and support for the cultural and economic development of their peoples and communities (Aylwin, 2000).

The 1993 Indigenous Peoples Law (Ley Indígena) establishes rules and principles for the protection, promotion and development of indigenous peoples, creating the National Corporation for Indigenous Development (Corporación Nacional de Desarrollo Indígena, CONADI). Under this law, nine "ethnic groups" are recognized, "the Mapuche, Aimara, Rapa Nui or Pascuenses, from the Atacameñas, Quechuas, Collas and Diaguita communities in the northern region of the country, the Kawashkar or Alacalufe and Yámana or Yagán from the southern channels" (Agostini *et al.*, 2010). During the late 1990s, territorial demands were made by different Lafkenche communities (from the Mapudungun language, "People of the Sea") and Williche (from the Mapudungun, People of the South) of southern Chile, organized in the Territorial Identity Lafkenche Organization (Organización Identidad Territorial Lafkenche) which groups communities living in Lafken Mapu. These communities sought the legal recognition of traditional practices related to the coastal border, including through the exercise of economic, social and religious activities, as a vital part of their culture. This movement validated the existence of an old political institution, which produced the first territorial proposal based on the claims of coastal communities from Golfo de Arauca to the Aysén region (Boccara, 2002). On the other hand, because of the possibility of strengthening indigenous peoples' national policy without the need to introduce urgent constitutional changes for the recognition of indigenous peoples' rights, Chile signed a series of international instruments which served to shape the current. In order to safeguard ancestral practices and territorial conservation, and after 17 years of complex parliamentary discussions, the ILO Convention 169 on the Rights of Indigenous and Tribal Peoples in Independent Countries (Zelada and Park, 2013) and UN Declaration on the Rights of Indigenous Peoples (UNDRIP, 2007) were ratified by the National Congress. These agenda-setting milestones strongly influenced the creation of a national indigenous policy concerning coastal and marine border communities, which led to the enactment of Law 20,249 which in turn created Marine and Coastal Areas for Indigenous Peoples (MCAIP).

Marine and Coastal Areas for Indigenous Peoples (MCAIP)

Law 20,249 defines an MCAIP as a defined coastal marine space, where management is handed over to communities or indigenous associations whose members have exercised customary uses of such spaces. The main objective of an MCAIP is to safeguard customary uses in order to maintain traditions and sustainable use of resources. The limits of MCAIPs are determined by the surface needed to support the exercise of customary practices. For its part, management must ensure the conservation of natural resources and support the well-being of communities in accordance with a management strategy and fisheries plan elaborated in conformity with current legislation applicable to different uses and activities taking place in the area.

For the establishment of an MCAIP there is a standardized, six-stage process which involves different public institutions. The complexity of this process is the result of the participation of a wide range of state agents, including the Ministry of Economy, Promotion and Tourism; Sub-Secretariat of Fisheries and Aquaculture; Ministry of Defense (through the Sub-Secretariat for the Armed Forces); Ministry of Social Development; National Corporation for Indigenous Development; and National Commissions for the Development of Coastal Borders. Additionally, as providers of technical information, the General Directorate of Maritime Territory and Merchant Navy and the National Fisheries Service are also involved in the process. In parallel to these institutions, an Inter-Sectorial Commission is created with representatives from the National Corporation for Indigenous Development, Sub-Secretariat of Fisheries and Aquaculture, the Ministry of Defense (through the Sub-Secretariat for the Armed Forces) and the General Directorate of Maritime Territory and Merchant Navy, to evaluate the management and fisheries plans and progress reports. The estimated administrative process takes 24 months, according to law (SUBPESCA, 2014).

The administrative process provides that within a maximum period of 14 months, as from the publication date of the application, an owner must present to the Sub-Secretariat of Fisheries and Aquaculture, a management plan describing the uses and activities to be undertaken in the MCAIP. This management plan contains the principles governing and objectives foreseen for the coastal area. It is basically the conceptual and operational structure reflecting activities to be undertaken, within the framework of natural, sociocultural and institutional realities, and the territorial dynamics influencing the area or space. These plans should contain: i) uses and frequency, ii) additional users (non-owners) whose activities and entry mechanisms are foreseen (can be any type of non-indigenous user and as a natural or legal person), iii) statutes of the community or indigenous association that incorporate internal norms and rules for conflict resolution, iv) program for dissemination activities, and v) submission of activity reports. In the case of the exploitation of hydrobiological resources, the plan shall include a fisheries management plan which ensures the effective conservation of resources and their environment. It must describe actions that will enable management of one or more fisheries based on updated knowledge regarding the bio-fishery, ecological, economic and social aspects affecting their existence, allowing identification of principles for the conservation and safeguard of these resources.

The case of Mañihueico-Huinay MCAIP

The Mañihueico-Huinay MCAIP began its application in 2010, based on a petition by nine Williches indigenous communities of the Hualaihué commune. The Hualaihué coastal marine area is located in the Lakes Region in the so-called "Chiloé Interior Sea" where the sinking of the central valley begins and creates a physical geography where the landscape reflects both mountain ranges, sea elements and fiord systems (Rojas, 2006). Based on these geographical features, the identity of Mapuche communities has been formed. This identity includes a strong territoriality and attachment to the coastal marine complex which is the main source of subsistence for these communities. The area provides support to both artisanal fishing activities and the extraction of resources, apart from being a communication path to populated areas in Chiloé, Calbuco and Puerto Montt. Mapuche get their supplies from these cities and their products are likewise traded in them until the present (Olea and Román, 2017).

In 1982, the first centers for salmonids production were established in these areas, triggering multidimensional transformations and, as a consequence, altering the territories and spatial, geographical and social configurations (Román, 2012; Olea and Román, 2017). The arrival of the salmon industry brought a number of changes, expressed by the significant increase of job offers and subsequent demographic growth that led to a rapid population increase of 42 percent between 1982 and 2017 (INE, 2018). This caused an important change in the working logic of the population, who went from a subsistence economy and autonomous management scheme, to a fully capitalist, competitive economy, where the use of new technologies is encouraged (Montero, 2004; Muñoz-Goma, 2009; Román 2012). Hand-in-hand with these strong transformations there was a failure to implement planning and land management processes, thus generating socio-spatial conflicts (Olea and Román, 2017). Delving deeper into these changes, neither of the two existing planning instruments of the commune – the Communal Development Plan and Regulation Communal Plan – addressed the coastal border, which explained the sociocultural and environmental consequences produced by the arrival of the salmon industry to the area. The real conflict in Hualaihué is the result of a new organization of the territory, based on modernization of economic activities, mainly aquaculture, exacerbating sociocultural changes among the local people and producing a dramatic transformation in the coastal border ecosystem (Olea and Román, 2017). This transformation is reflected in increased pressure over the natural resources of coastal areas, including from environmental impacts caused by the salmon industry, expressed in the appearance of infectious salmon anemia since 2005, that affected marine resources and the population's socioeconomic situation; greater presence of waste in the coastal and marine space associated with the aquaculture industry; as well as consequences generated on the seabed due to stresses from aquaculture (Buschmann and Fortt, 2005; Pinto, 2007; Buschmann *et al.*, 2009).

The Mañihueico-Huinay MCAIP was approved in 2018 by the Regional Commissions for the Use of Coastal Borders and appears as a new management tool, joining a set of existing instruments for coastal borders. In other words, the incorporation of the MCAIP into the territorial planning system ratifies the inclusion of a new approach and look which safeguards the traditional use of coastal marine resources and promotes an integrated relationship (Olea and Román, 2017). Through the Association of Indigenous Communities of the Walaywe Territory, the Mañihueico-Huinay MCAIP defines its own governing, planning and spatial management principles over 910 km^2 declared on the coastal marine border (Figure 8.1). The Association promotes support and guidance to develop collective capacities that facilitate social learning of current norms that affect the coastal space and indigenous peoples.

As a result, since the request for an MCAIP, considerable empowerment of Association leaders has been observed. These leaders began to foresee potential conflicts and socio-environmental problems that could be addressed through MCAIPs. Among the most severe conflicts, the salmon industry and collective rejection of its bad practices that affect both the coastal border and the seabed have been highlighted (Yohana Coñuecar, personal communication, October 17, 2018). Together with these problems, new threatening scenarios have been identified, including the presence of mining exploration and violations of the right to water. The Association has also identified the need to actively safeguard wetlands as priority sites for subsistence. The gradual strengthening of the Association and growing capacity of its leaders has resulted in a new political player which, through the implementation of the MCAIP, is seeking to address social demands with regards to the distribution, recognition and management of ecosystem services that the coastal border territory offers.

At the same time, productive sectors in large-scale aquaculture see the MCAIPs as a threat as their creation reduces the availability of otherwise free marine-coastal areas to cultivate, and uses of these areas are granted exclusively to MCAIP holders. Often these uses are not compatible with the needs of the aquaculture sector. Additionally, an application for the creation of an

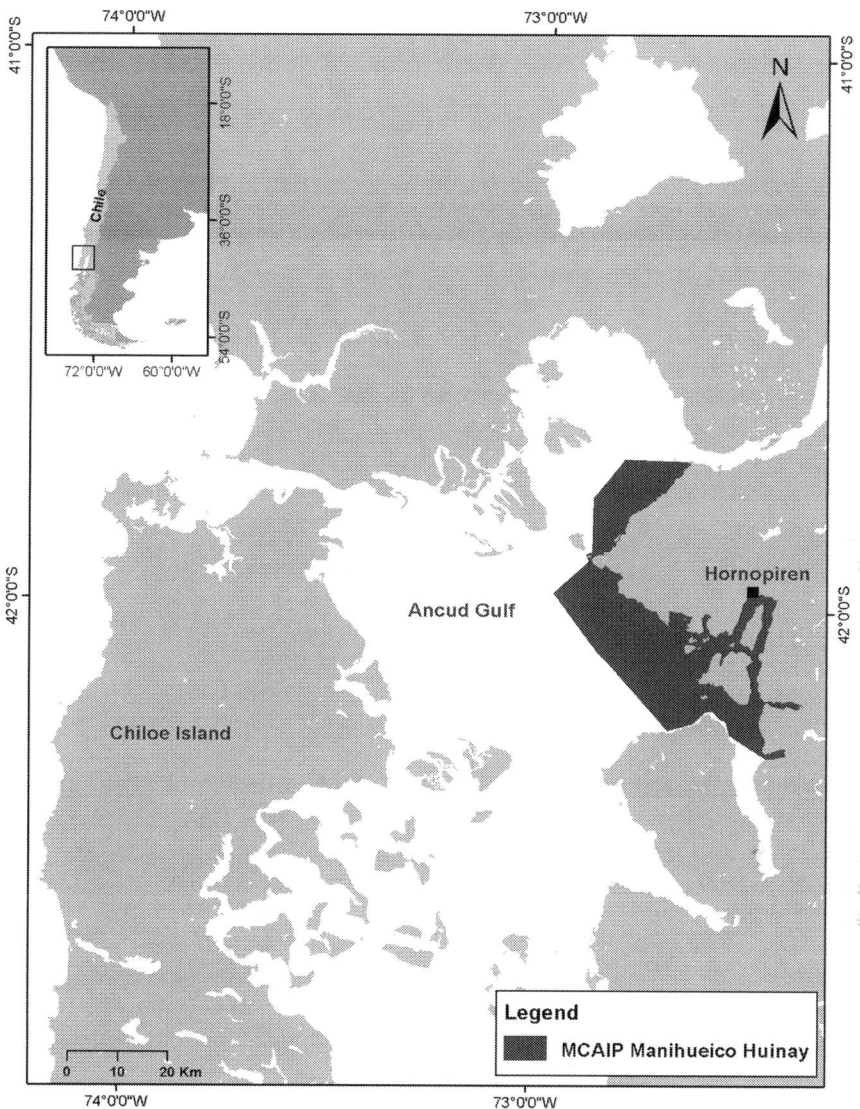

Figure 8.1 Geographic location representing 910 km² of the Marine and Coastal Areas of the Mañihueico Huinay Indigenous Peoples.

MCAIP implies the suspension or rejection of all other applications which may affect coastal border uses. In the light of growing legal uncertainties in this sector, the Association convened a discussion table, with the participation of non-indigenous fishermen, mytilidae producers, salmon producers, tourist groups and other indigenous communities living in the area. The discussion table operated based on a set of rules established by the Association. Three representatives of each sector would participate in monthly meetings. During the operations of the discussion table, this has had the effect of positioning the Mañihueico-Huinay MCAIP as an administrative body, capable of triggering collective actions and allowing the establishment of communication channels and coordinated actions among different actors in a localized space. It has also permitted different forms of territorial conflict resolution regarding the coastal area to be conserved (Gelcich *et al.*, 2015). This discussion table helped to conciliate different interests of different coastal border users, particularly with regards to environmental and sociocultural aspects, at the same time having the effect of recognizing customary practices, uses and rights of indigenous peoples over coastal and marine areas (Andrade, 2000).

The Mañihueico-Huinay MCAIP showcases a political dimension and action by the Association of Indigenous Communities of the Walaywe, which was capable of deploying a negotiating strategy with government institutions and private institutions to ensure binding agreements. As a result of their action and pressure, the National Corporation for Indigenous Development developed a second report on customary uses that validated ancestral practices of communities and the Regional Commissions for the Use of Coastal Borders acknowledged exclusions and the agreements reached at the discussion table, regarding the cartographic map of the MCAIP. Furthermore, a claim by the Association to the National Controller Office accelerated voting on the future of the Mañihueico-Huinay MCAIP in the Regional Commissions for the Use of Coastal Borders.

In retrospect, the Hualaihué process has made it possible to foresee the potential and opportunities of an MCAIP as a territorial planning instrument, in addition to acting as a catalyst for negotiating processes and conflict resolution among relevant actors. The Mañihueico-Huinay MCAIP case demonstrates that this planning instrument is capable of supporting integrated management of a defined territory, where the importance of ancestral uses in the coastal border is underscored and plays a critical role. The MCAIP also shows how technical-traditional knowledge of local indigenous peoples and communities can be a starting point for land-use planning and strengthen organizational, participation and dialogue capacities among parties involved in the development and conservation of the coastal and marine space.

Final reflections

In addition to existing instruments which recognize rights and access to coastal and marine management of areas and resources in Chile, new instruments have been developed in favor of coastal communities, mainly triggered by the adoption of international instruments such as in the ILO Convention 169 and the UNDRIP. The enactment of the law to create MCAIPs represents a major advancement in the recognition of collective rights of indigenous peoples, particularly given its binding legal nature. In political terms, it creates a territorial management tool that enables communities to organize their territories based on autonomy, self-management and their ancestral institutions and practices. MCAIPs seek to safeguard specific spaces in coastal and marine areas, resources and the rights of local populations therein.

With regards to fisheries law and rights, it can safely be argued that internationally, there has been a move to recognize the right to access and use of natural resources by indigenous peoples (Ban *et al.*, 2008). Though to some extent Chile has followed this trend, particularly with the implementation of the of the General Law on Fisheries and Aquaculture and its amendments, conflict has arisen and focused on concerns by coastal indigenous communities, regarding the lack of provisions that specifically recognize and guarantee their access rights and rights to fisheries resources (FAO, 2016). However, and in parallel, reports on customary use prepared by the National Corporation for Indigenous Development have approved ancestral practices of the MCAIP and therefore validated fishing and harvesting activities, under the access and/or management regime regulated under the General Law on Fisheries and Aquaculture.

Generally, coastal indigenous communities that depend on natural resources management implement a truly ecosystem-based approach which integrates economic, spiritual, cultural, resource and environmental dimensions (Outeiro *et al.*, 2015; Berkes, 2004). The establishment and implementation of MCAIPs produces a co-management regime for coastal and marine areas. They provide communities with an opportunity to self-determine their own governance systems under an approach founded on their indigenous knowledge, experiences, world-views and institutions (Rugebret, 2015). In similar cases, indigenous communities have maintained control over coastal and marine areas and their resources under traditional management and customary practices reflected in agreements which guide and orient correct and sustainable use of their resources (Butler *et al.*, 2012; Cinner *et al.*, 2012; Davies *et al.*, 2013; Nursey-Bray and Jacobson, 2014; Rugebret, 2015).

Even though the creation of MCAIPs recognizes and validates customary uses by indigenous peoples, they have not been exempt from controversy. Since the General Law of Fisheries and Aquaculture entered into force, aquaculture productive sectors – mainly the salmon sector – have been critical of the attributes and recognition of spaces provided by the law. This is primarily as a result of "freezing" other requests and applications for activities in the marine space covered by the MCAIP. Additionally, given the existence of a tight race to gain rights over a portion of the coastal and marine border in Chile, coastal indigenous communities have started to take center stage with their strong actions to safeguard the natural patrimony and ensure traditional activities, such as artisanal fishing and protect religious, cultural and medicinal prayer zones. These actions ensure spaces for the development of local practices that would improve the well-being of coastal communities in general. Although conflict has arisen, the MCAIP has in turn served as an instrument to address and mitigate its effects. The creation of discussion tables among multiple actors, for example, and the incorporation of non-indigenous actors, have helped to placate some of the problems and issues affecting other stakeholders with converging interests in these spaces and areas. As a result of their precautionary basis and the fact that they respond to local realities, the MCAIP has become a robust instrument to foster local and community empowerment.

The establishment of MCAIPs generates opportunities to revitalize cultural aspects of a territory together with empowering communities in terms of valuing their knowledge, practices and their customs. The rescue of ancestral knowledge and practices is crucial, particularly as conservation strategies require multiple and culturally sensitive approaches. However, there are still important challenges to be overcome in these biodiversity-rich areas, even when positive results for conservation are observed under the implementation of MCAIPs (Berkes, 2009). Internationally, the establishment and creation of similar conservation areas and/or planning instruments has been evolving and rapid over the last decades. For instance, Indigenous and Community Conserved Areas/ICCA), are defined in general as:

> natural and/or modified ecosystems containing significant biodiversity values and ecological services, voluntarily conserved by (sedentary and mobile) indigenous and local communities, through customary laws or other effective means.
>
> (IUCN Durban World Park Congress 2003)

MCAIPs would "fit" under this general category – although mostly focused on conservation. These spaces and areas are characterized by i) involving one or more communities that have a close connection with the ecosystem; ii) involving management actions by the community which contribute to conservation; and iii) recognizing that the community is the main actor in decision-making (Berkes, 2009). It is possible to illustrate the potential of MCAIPs to biodiversity conservation in particular and the recognition of traditional knowledge for the management of natural resources (Raymond *et al.*, 2010), their contribution to resilience (Gadgil *et al.*, 1993) and environmental adaptation (Folke, 2016). Finally, as a result of the law which created MCAIPs, key opportunities and challenges are present to meaningfully face the management of coastal and marine ecosystems, recognize customary rights, support self-determination of communities over their territories and, ultimately, contribute to strengthening culture, conservation and sustainable development of coastal and marine spaces.

Acknowledgements

We wish to thank The Walton Family Foundation for their contribution to this chapter. We would also like to thank the Manquemapu, Mahuidantu, San Pedro and Pu Wapi indigenous communities and the Association of Indigenous Peoples of Mañihueico-Huinay, for their dedication, commitment and confidence in our work. Finally, we thank Yohana Cuñuecar for her contributions to improve part of this manuscript.

Notes

1 According to Stavenhagen, 1990, customary law refers to a custom, as a source for the construction of norms within a certain socio-cultural context.
2 Chile is ranked ninth in terms of the number of environmental and social conflicts according to the OECD (2016).

References

Agostini, C. A., Brown, P. H., Roman, A. C. (2010), Poverty and Inequality Among Ethnic Groups in Chile. *World Development, 38*(7), 1036–1046.

Andrade, B. (2000), Los Espacios Litorales: Definiciones, Actores, Desafíos, Perspectivas. In F. Arenas, & G. Cáceres, *Ordenamiento del Territorio en Chile: Desafíos y urgencias para el tercer milenio*. Ediciones Universidad Católica de Chile, Santiago de Chile, 21–30.

Aylwin, J. (2000), Los Conflictos en el Territorio Mapuche: Antecedentes y Perspectivas. *Perspectivas, 3*(2), 277–300.

Ban, N. C., Picard, C., Vincent, A. C. J. (2008), Moving Toward Spatial Solutions in Marine Conservation with Indigenous Communities. *Ecology and Society, 13*(1), 32. Available www.ecologyandsociety.org/vol13/iss1/art32/.

Barragan, J. M. (2003), *Medio Ambiente y Desarrollo en Areas Litorales*. Servicio de Publicaciones de la Universidad de Cádiz, España.

Berkes, F. (2004), Rethinking Community-Based Conservation. *Conservation Biology, 18*(3), 621–630.

Berkes, F. (2009), Community Conserved Areas: Policy Issues in Historic and Contemporary Context. *Conservation Letters, 2*(1), 19–24.

Boccara, G. (2002), The Mapuche People in Post-Dictatorship Chile. *Estude Rurales, 163–164*, 283–304.

Boccara, G., Seguel-Boccara, I. (1999), Políticas Indígenas en Chile (siglos XIX y XX). De La Asimilación al Pluralismo (el caso mapuche). *Revista de Indias*. 59(217), 741–774.

Buschmann, A., Fortt, A. (2005), Efectos Ambientales de la Acuicultura Intensiva y Alternativas para un Desarrollo Sustentable. *Revista Ambiente y Desarrollo, 21*(3), 58–64.

Buschmann, A., Cabello, F., Young, K., Carvajal, J., Varela, D., Henríquez, L. (2009), Salmon Aquaculture and Coastal Ecosystem Health in Chile: Analysis of Regulations, Environmental Impacts and Bioremediation Systems. *Ocean & Coastal Management*, 52(5) 243–249.

Butler, J. R. A., Tawake, A., Skewes, T., Tawake, L., McGrath, V. (2012), Integrating Traditional Ecological Knowledge and Fisheries Management in the Torres Strait, Australia: the Catalytic Role of Turtles and Dugong as Cultural Keystone Species. *Ecology and Society, 17*(4), 34. http://dx.doi.org/10.5751/ES-05165-170434.

Cinner, J. E., Basurto, X., Fidelman, P., Kuange, J., Lahari, R., Mukminin, A. (2012), Institutional Designs of Customary Fisheries Management Arrangements in Indonesia, Papua New Guinea, and Mexico. *Marine Policy, 36*(1), 278–285.

Davies, J., Hill, R., Walsh, F. J., Sandford, M., Smyth, D., Holmes, M. C. (2013), Innovation in Management Plans for Community Conserved Areas: Experiences from Australian Indigenous Protected Areas. *Ecology and Society*, *18*(2), 14. http://dx.doi.org/10.5751/ES-05404-180214.

Duquesnoy, M. (2012), The Tragedy of the Utopia of the Mapuche of Chile: Territorial Vindications in the Times of Applied Neoliberalism. *Revista Paz y Conflictos*, *5*, 20–43.

FAO. (2016), *Informe Final Proyecto Asistencia para la Revisión de la Ley General de Pesca y Acuicultura, en el Marco de los Instrumentos, Acuerdos y Buenas Prácticas Internacionales para la Sustentabilidad y Buena Gobernanza del Sector Pesquero* (UTF/CHI/042/CHI).

Folke, C. (2016), Resilience (Republished). *Ecology and Society*, *21*(4), 44. https://doi.org/10.5751/ES-09088-210444.

Gadgil, M., Berkes, F., Folke, C. (1993), Indigenous Knowledge for Biodiversity Conservation. *Ambio*, *22*(2/3), 151–156.

Gelcich, S., Peralta, L., Donlan, J., Godoy, N., Ortiz, V., Tapia-Lewin, S., et al. (2015). Alternative Strategies for Scaling up Marine Coastal Biodiversity Conservation in Chile. *Maritime Studies*.

Instituto Nacional de Estadística. (19 October 2018), *Resultados Censo 2017. Por País, Regiones y Comunas*. Available at http://resultados.censo2017.cl/.

IUCN. Durban World Park Congress (2003), WPC Recommendation 26. Community Based Conservation. Available at, http://danadeclaration.org/pdf/recommendations26eng.pdf.

Miller, R. (2012), The International Law of Colonialism: A Comparative Analysis. *Lewis & Clark Law Review*, *15*(4), 871.

Montero, C. (2004), *Formación y Desarrollo de un Clúster Globalizado: el Caso de la Industria del Salmón en Chile*. Santiago de Chile: Serie Desarrollo Productivo – CEPAL.

Muñoz-Goma, O. (2009), *Aguas Arriba: Transformación Socioeconómica del Ecosistema Llanquihue-Chiloé (Chile) Durante los Años 90*. Buenos Aires: CLACSO.

Nursey-Bray, M., Jacobson, C. (2014), Which way?: The Contribution of Indigenous Marine Governance. *Australian Journal of Maritime & Ocean Affairs*, *6*(1), 27–40.

OECD Environmental Performance Reviews: Chile (2016), Available at, www.oecd.org/chile/oecd-environmental-performance-reviews-chile-2016-9789264252615-en.htm.

Olea, J., Román, J. (2017), El Ordenamiento Territorial y Modernización de la Patagonia Norte Chilena. El Caso de Estudio de la Comuna de Hualaihué: Borde Costero, Salmoneras y Comunidades Indígenas. *Planeo*. Available at www.researchgate.net/publication/313268542.

Outeiro, L., Gajardo, C., Oyarzo, H., Ther, F., Cornejo, P., Villasante, S., Ventine, L. B. (2015), Framing Local Ecological Knowledge to Value Marine Ecosystem Services for the Customary Sea Tenure of Aboriginal Communities in Southern Chile. *Ecosystem Services*, *16*, 354–364.

Pinto, F. (2007), *Salmonicultura chilena: Entre el Exito Comercial y la Insustentabilidad*. Santiago de Chile: Terram Publicaciones.

Raymond, C. M., Fazey, I., Reed, M. S., Stringer, L. C., Robinson, G. M., Evely, A. C. (2010), Integrating Local and Scientific Knowledge for Environmental Management. *Journal of Environmental Management*, *91*(8), 1766–1777.

Rojas, O. (2006), Caracterización de la Tectónica del Territorio Chileno. Concepción: Departamento de Geografía – Universidad de Concepción.

Román, J. (2012), *Hornopirén (1973–2007), Tres Décadas de Cambios, Contradicciones y Paradoja. Análisis Histórico del Proceso de Desarrollo Económico Capitalista en "Chiloé Continental"*. Faculty of Philosophy and Humanities: University of Chile. Final Report to Obtain a Degree in History.

Rugebret, R. V. (2015), The Environmental Management Philosophy of Indigenous People in Coastal Marine Area in Maluku. *International Journal of Advanced Research*, 3(7), 1322–1329.

Schlotfeldt, C. (2000). Consideraciones metodológicas y conceptuales para el ordenamiento costero: análisis de estudio de caso. In F. Arenas & G. Cáceres, *Ordenamiento del Territorio de Chile: Desafíos y urgencias para el tercer milenio*. Ediciones Universidad Católica de Chile, Santiago de Chile, 39–58.

Serani, J. (2013), Zonificación del Borde Costero: Turismo y Recreación. En Bermúdez y Hervé Editores: *Justicia Ambiental. Derecho e Instrumentos de Gestión del Espacio Marino Costero*. Ediciones Pontificia Universidad Católica de Valparaíso, pp. 429–459.

Stavenhagen, R. (1990), *Entre la Ley y la Costumbre*, México, Instituto Indigenista Interamericano/Instituto Interamericano de Derechos Humanos.

SUBPESCA. (2014), *Guía para la Aplicación de la Ley de Espacios Costeros Marinos para Pueblos Originarios*. Ministerio de Economía, Fomento Y Turismo. Subsecretarpía de Pesca Y Acuicultura, 45.

Zelada, S., Park, J. (2013), Análisis Crítico de la Ley Lafkenche (No. 20.249). El Complejo Contexto Ideológico, Legal, Administrativo y Social que Dificulta su Aplicación. *Universum*, 1, 47–72.

9 Protection of migratory marine species

Ximena Vélez-Zuazo

Introduction

Movement between distant habitats is probably the most distinct feature of marine migratory species. At the same time, this presents major challenges to implement successful conservation measures. During their lifetime, these species travel through different natural spaces sometimes separated by thousands of kilometers, crossing complete oceans and continents, regardless of the geopolitical limits or areas beyond jurisdiction (Harrison *et al.*, 2018). As far as records go, no species comes even close to the Arctic tern (*Sterna paradisaea*) that over a lifetime flies the equivalent of three trips to the moon … and back (Egevang *et al.*, 2010)!

Migration can respond to different needs. One of them is finding new feeding and growing grounds as in the case of the hawksbill turtle or Southern right whale (Valenzuela *et al.*, 2009; Velez-Zuazo *et al.*, 2008), as well as breeding and nesting grounds as in the case of humpback whales (Acevedo *et al.*, 2017). They can also temporarily migrate as a response to temperature changes (winter/summer), lack of food or to continue their development (neonates/juveniles/adults). The habitats they occupy respond to needs of the moment and can be dissimilar. For example, when marine turtles are born, they abandon their nests in the sand and head towards open seas in search of macroalgae patches. Later they temporarily establish in coastal habitats – like mangroves, reefs or sea grasses – and when they reach maturity, reproductive migration begins opposite the sand beaches where they were born. Throughout this process, migratory marine species provide important ecosystem services, including regulation services (i.e. supply of nutrients through excretion), provisioning (i.e. guano harvesting) and cultural (i.e. recreational experiences) (Reynolds and Clay 2011; López-Hoffman *et al.*, 2017).

Although animal migration has been studied at an increasing rate, the same cannot be said about species that migrate extensively. This continues to be a major challenge due to the temporal and spatial complexity that characterizes the movement of migratory animals. The challenge is even greater if interaction with human activities is added. Many species have suffered dramatic reductions in their populations as a result of conflicts with humans, the impact of fisheries activities, developments on coastal areas and marine pollution. A reduction of their populations has direct negative consequences for the environment (i.e. reduction of top predators) (Heithaus *et al.*, 2012) and can trigger unexpected events. To mitigate the impact of human activities on migratory species, international instruments have been developed which countries have joined over time. However, effective implementation of these agreements depends on each country.

In the Eastern Pacific, Chile, Mexico and Peru have an important fishery industry with common challenges for the conservation of migratory marine species. They also share common species in their seas. This chapter will address the common threats for these three countries, the international agreements to which they are parties and the existing opportunities to improve the protection of migratory marine species that move through their territories. As a case study, we analyze the state of conservation of sea turtles in the Eastern Pacific.

Geopolitics of marine species migration

One of the studies that better summarizes the transnational nature of marine species movements is by Harrison *et al.*, 2018 (Figure 9.1). He presents a compilation of migratory routes obtained through satellite tracking of 1,648 individuals of 14 different species of large marine vertebrates. The results of the study focused on populations in the North Pacific Ocean and shows the magnitude of marine species movements. The individuals monitored travelled thousands of kilometers and occupied the jurisdictional spaces (Exclusive Economic Zones) of 37 different countries. The origin and routes, distances travelled and time invested in their migration is as diverse as the target species in this study.

In the aforementioned study as well as in previous publications on migratory species, four important aspects should be considered as part of the geopolitics of species migration: (1) the connection established between jurisdictions during their lifetime, (2) the number of different jurisdictions crossed during migration, (3) the seasonal pattern of visits in each jurisdiction, and (4) the time they remain beyond any national jurisdiction. This information is key to analyze the effectiveness of protection policies, to identify legal gaps and the necessary changes and/or adjustments required, and what policies need to be developed and implemented. The study of marine migratory species movements provides evidence on the variety of movement patterns and geopolitics of their journeys. From the study by Harrison *et al.*, it can be observed that during their migration species occupied areas under clear jurisdiction as in the case of the long-tailed jaeger (*Stercorarius longicaudus*), which during their migrations in non-reproductive periods visited exclusive economic zones in various countries including Chile, Mexico and Peru (Figure 9.2). In contrast, leatherback sea turtles (*Dermochelys coriacea*) — that nest on protected beaches in Central America — spend most of their lives in open waters, representing a huge challenge for conservation, taking into

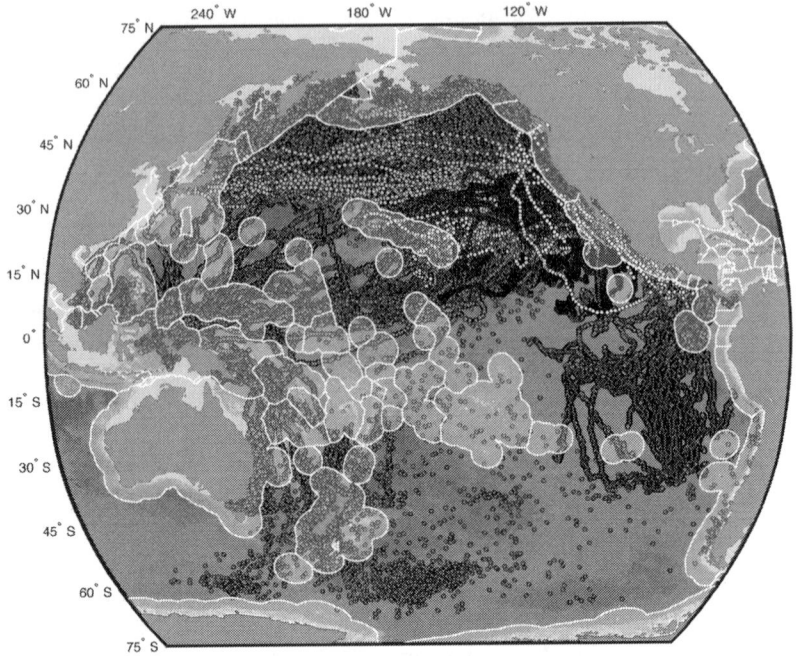

● Albacore tuna   ○ Pacific bluefin tuna   ● Yellowfin tuna   ● Blue shark   ● Shortfin mako shark
● Salmon shark   ○ White shark   ● Leatherback turtle   ● California sea lion   ○ Blue whale
◎ Northern elephant seal   ● Black-footed albatross   ● Laysan albatross   ● Sooty shearwater

Figure 9.1 Migratory routes drawn from the study of 14 marine species between 2000
and 2009 in the North Pacific Ocean and the jurisdictional spaces occupied (translucent white polygons) during that time. Each circle along the
route indicates the daily location.

(Courtesy of Autumn-Lynn Harrison and *Nature Ecology & Evolution*.)

consideration that they remain for a long time in spaces beyond jurisdiction
or unprotected zones. In fact, populations of the Eastern Pacific are nowadays
critically endangered due to the problematic interactions with high-seas fisheries. As with cormorant populations, protection instruments should reach all
migratory marine species, including those with shorter dispersal distances of
less than 10 kilometers. In these cases, conservation actions must be more
focused and supported by national protection polices (Oppel *et al.*, 2018).

These two examples summarize the complexity of implementing migratory
species protection measures, which was detected early in shorebirds, the most
difficult migratory species to track with satellite sensors due to their small size
compared to mega-vertebrates. Before the development of technologies to
monitor marine species with large devices in real time, it was already evident
that migratory species covered long distances. A great amount of information
has come from studying the movement habits of shorebirds using bird banding,
as they travel thousands of kilometers to reach their reproduction or winter

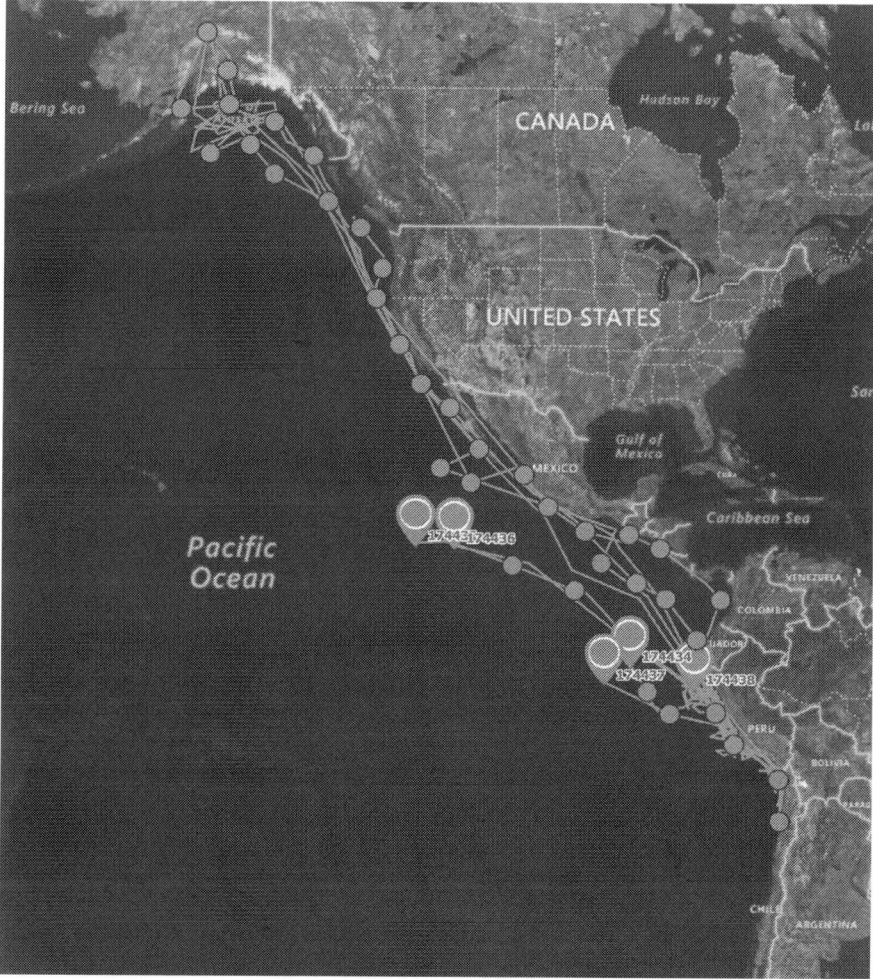

Figure 9.2 Migration of the long–tailed jaeger (*Stercorarius longicaudus*) between its reproduction areas in Alaska and non–reproduction areas in various countries in the Southern Hemisphere and South America.

(Courtesy of Autumn-Lynn Harrison.)

sites (Pitelka, 1979). It is also clear that their life history, strongly linked to migration, represented a risk well before the conservation status of these species was a concern for science (Myers *et al.*, 1987).

For the effective conservation of migratory species this scenario demands policy attention from three levels. Some are for migratory species in general, others are focused on groups of specific species. The first level involves international policies such as the Convention for the Conservation of Migratory Species (CMS). A second level involves regional agreements, such as the

Inter-American Convention for the Protection and Conservation of Sea Turtles. Finally, a third level relates to policies adopted by countries, such as in the case of Peru which enacted a law that prohibits the extraction of whale sharks.[1]

At the same time, for conservation policies on migratory marine species shared by Chile, Mexico and Peru to be successful, a joint strategy is required. This means adherence of these countries to international and regional agreements, as well as the development and implementation of inter-sectorial policies. One advantage for these countries is that they share a strong and important fisheries culture and similar conservation challenges to mitigate the many threats migratory species are faced with during their life.

Threats to the conservation of migratory marine species in the Eastern Pacific

Migratory marine species face a number of threats throughout their life history. These can be very specific, referring to immediate effects (i.e. volcanic erup-tions, tsunamis, etc.), or gradual, those that increase in intensity (i.e. climate change, algae overgrowth and habitat reduction (Sutherland *et al.* 2012). In a study published by Sutherland *et al.*, in 2012, an extensive list of identified threats for nearly 200 species of shorebirds was compiled. Two years later, Las-celles *et al.*, (2014) contributed with research on a list of threats faced by the mega-vertebrates. This research identifies the ten most important threats affect-ing species on the Red List of Threatened Species of the International Union for Conservation of Nature (IUCN). Fisheries, invasive species, pollution and climate change compromise the existence of 1,666 species (82.9 percent of species evaluated). The data provides evidence about the three main survival threats to migratory marine species in Mexico, Chile and Peru: bycatch, loss of habitat and pollution of coastal and marine environments.

Bycatch

This directly impacts non-target species during fishing operations and is one of the most serious threats to migratory marine species (Žydelis *et al.*, 2013). Bycatch occurs both in jurisdictional waters and the high seas, marine spaces over which no country exercises jurisdiction. This problem is a result of migratory routes for marine species overlapping with fishing areas (Lewinson *et al.*, 2004) and the use of fishing gear and/or strategies that only concentrate on effectiveness in fishing of target species, without consideration of impacts on other species that inhabit or pass through the same ecosystem. In the Mexican Pacific, artisanal fishing is primarily responsible for bycatch of log-gerhead sea turtles (*Caretta caretta*). This causes the death of 1,000 turtles in a single season (Peckham *et al.*, 2007) with far-reaching effects on large marine mammals (Medellin *et al.*, 2009). In Peru, the effects of bycatch have been evaluated in different types of fishing activities (nets and longline) and for various species. In general, artisanal fishing is considered to cause major impacts on populations of marine migratory species. Bycatch of albatross (Awkerman *et al.*, 2006), sea turtles (Alfaro–Shigueto *et al.*, 2011), sharks (Gilman *et al.*, 2008) and marine mammals (Mangel *et al.*, 2010; García-Godos *et al.*, 2013) has been reported and estimated. In Chile, the same problem has been reported for over 27 sea bird species (Suazo *et al.*, 2014).

Reduction and/or loss of critical habitat

The movement of migratory species takes place between certain areas, responding to specific roles in their lifecycle. For example, some species such as salmon and sea turtles return time and again for their reproduction to the same areas where they were born, a life trait called natal philopatry (Dittman *et al.*, 1996, Meylan *et al.*, 1990). Something similar happens with various migratory marine species, accustomed to visiting the same wintering areas for food and to develop, and the same rest sites where they return throughout their lives. All these areas identified as natural habitats (i.e. wetlands, river mouths, coral reefs, sand beaches, desserts and marine prairies) and modified (i.e. rice fields), have unique characteristics and are critical spaces to ensure the migration of these species. The modification, reduction and, in some cases, the disappearance of these areas, have damaging effects for conservation of species and causes a decrease of migratory birds (Robbins *et al.*, 1989; Sauer *et al.*, 2005). More than half the bird species using well-established routes called "Flyways" have decreased in abundance due to the disappearance of critical rest areas that used to exist throughout their journeys. Additionally, Kirby *et al.*, (2008) has identified agriculture, aquaculture, the exploitation of resources and transformation of natural systems (i.e. construction of dams or drained wetlands) as activities which produce negative impacts. In short, habitat loss and degradation are two of the main threats that affects migratory species, as areas used throughout their migration are often not protected in any way (i.e. shorebirds) (Runge *et al.*, 2015). In Mexico, the loss of habitat due to aquaculture activities and developments on the coastal strip, are responsible for the decline of the snowy plover (*Charadrius nivosus*) (Cruz-López *et al.*, 2017) and populations of the American oystercatcher (*Haematopus palliates frazari*) (Palacios *et al.*, 2017); while in Peru, large extensions of wetlands have been lost, covering kilometers of critical habitat for various species of migratory birds (i.e. Pantanos de Villa) (Pulido and Bermúdez 2018).

Pollution

Pollution of aquatic environments also puts conservation of migratory species at risk, according to the Red List of Threatened Species (Lascelles *et al.*, 2014). The problem can originate from different sources, happen in varying magnitudes and depending on the characteristics, generate an impact on populations and their reproductive success. Pollution can be produced by oil spills (i.e. the Deepwater Horizon spill) (Beyer *et al.*, 2016), urban and industrial discharges, plastic pollution (Derraik, 2002) and even acoustic pollution (Graham *et al.*, 2017). In Mexico, for example, pesticide presence in the tissue of three species of sea turtles has been detected (Juarez, 2004). In Chile, the effect of sound has been studied in marine mammals. Ribeiro *et al.* (2005) found that the Chilean dolphin (*Cephalorhynchus eutropia*) modifed their behaviors in the presence of vessels. Pollution from plastics has also been studied in Peru and Chile (Thiel *et al.*, 2011). Chile has notified the presence of these residues in remote marine zones (Pérez-Venegas *et al.*, 2017) and in areas occupied by various bird species (Miranda-Urbina *et al.*, 2015). Micro-plastics, to be more specific, have been identified as a potential threat for migratory birds (Cole *et al.*, 2011; Sutherland *et al.*, 2012).

As a result of the accumulated effect of all these threats until 2014, 21 percent of migratory marine species now appear on the IUCN Red List of Threatened Species.

The conservation of migratory species through a case study: sea turtles

Within the group of migratory species, sea turtles encompass the complexity of natural history and challenges for the conservation of high mobility species. As a group of well-studied species, they are a good case for analysis. In a simple way, sea turtles are born on sandy beaches and make their way to open waters where they spend most of their life. After a few years, some proceed to shallower areas while others continue in open waters, their feeding grounds and growing habitats. Sea turtles will faithfully remain in these areas until their reproductive age, and upon reaching maturity, will swim back to the beach on which they were born to lay eggs. During their adult history, sea turtles migrate again and again between feeding grounds and reproduction/nesting areas (see Figure 9.3).

Their affinity to move between distant areas ensures them the resources for reproduction and growing, but also exposes them to different threats (Figure 9.3). Today, conservation instruments are extremely diverse and numerous. Frazier (2014) compiles a complete list of binding international and regional instruments for countries in the South Eastern Pacific, including Peru and Chile. An excerpt of the list is presented in Table 9.1, with the threats they face. Two regional instruments have been considered – the CPPS and the Lima Convention – developed and implemented as part of the Regional Program for the Conservation of Sea Turtles of the South Eastern Pacific. This program also addresses the best scenario for countries to collaborate in sea turtle conservation. The strength of this conservation instrument, as explained by Frazier (2014), lies in the similarity between countries, collaboration and repeated commitments by signatory countries, as well as in different international and regional instruments and the program's objectives.

At first glance, these agreements address the most important threats and produce links between countries. However, concrete actions continue to be the responsibility of each country. Each country has taken actions to mitigate

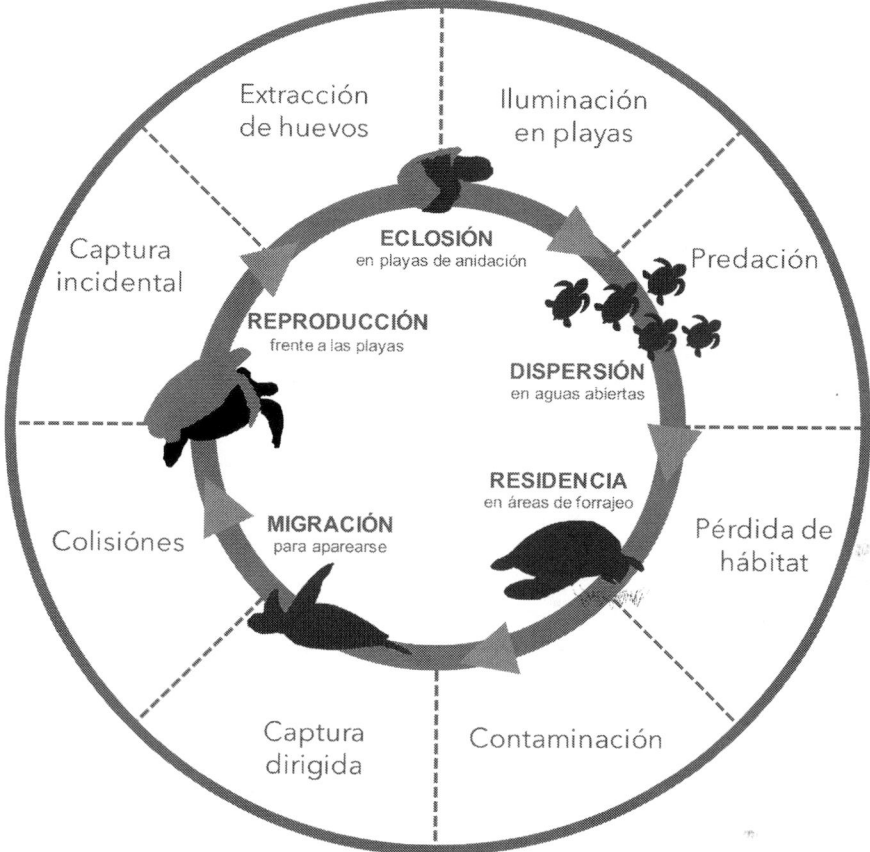

Figure 9.3 Generalized lifecycle of a sea turtle and the threats they face. Some threats persist throughout their lives while others are exclusive to some stages in their life history.

Table 9.1 Some existing international and regional instruments adopted by Peru and/ or Chile, and the main threats they face to protect sea turtles. Extracted and modified from Frazier (2014)

Scope	Instrument	Threat addressed	Mexico	Peru	Chile
International	CITES	International trade	X	X	X
	CMS	Loss of habitat, by-catch	–	X	X
	MARPOL	Marine contamination	X	X	X
	World heritage	Loss of habitat	X	X	X
	Ramsar	Loss of habitat	X	X	X
	CONVEMAR	Various	X	X	–
Regional	Convenio de Lima	Loss of habitat	–	X	X
	CIAT	By-catch	X	–	X
	IAC	Various	X	X	X
	CPPS	Loss of habitat	–	X	X

the threats for these groups of species at a local level.[2] These vary among each one of them and include indefinite fishing bans, the implementation of bycatch reduction devices, best practices on board vessels with bycatch events (i.e. recovery and release), the restriction of fisheries activities in areas adjacent to critical habitats, among others. The adhesion to the Inter-American Convention for the Protection and Conservation of Sea Turtles, plays a central role as it promotes the implementation of a National Action Plan at the local level. These plans produce close collaboration between government and civil society. Until 2018, according to national reports presented to the Inter-American Convention, only Mexico has developed a conservation plan, as part of an Endangered Species Recovery Plan.[3] In Peru, the Forestry Service leads the drafting of a proposal for the National Plan for Conservation of Sea Turtles. The last working meeting took place in April 2018, where the final proposal was consolidated and approved for validation by local actors. In 2018, Chile was still working on a National Action Plan.

If the policies of the three countries are reviewed and compared, one can observe that although Mexico is ahead of Peru and Chile in the development of multiple local fisheries policies, including a local program for the conservation of sea turtles, Mexico has not adhered to the Convention on Migratory Species. On the other hand, although Peru and Chile have already joined the Convention, they have not finished the design and approval of their local conservation plan for these marine species.

Challenges and opportunities for the conservation of migratory marine species

Despite extensive work in the region regarding policies and programs for sea turtle conservation, populations in the Eastern Pacific continue to be threatened (Velez-Zuazo *et al.*, 2017). There is a considerable reduction of populations, exceeding 97 percent in some cases, as with the Leatherback Sea Turtle (*Dermochelys coriácea*). The continuous decline of this species is mainly due to bycatch in open waters and poaching of their eggs from nesting beaches, two critical habitats in their life history. The case of the Leatherback Sea Turtles in the Pacific is an example of how, despite efforts, the survival of migratory species is not guaranteed, and the mitigation of threats continues to be a challenge.

Identification of critical use areas

There are initiatives to reduce the interaction of migratory marine species with fisheries that are not exclusive for sea turtles. To date, effectiveness of different strategies to mitigate bycatch has been researched quite thoroughly. These strategies include direct communication with fishers during their fishing operations (Alfaro-Shigueto *et al.*, 2012), the replacement of "j" type fishing hooks with "offset" circle hooks in longline fishing (meta-analysis by Reinhardt *et al.*, 2018), the use of acoustic devices (Mangel *et al.*, 2013) and lighting (Ortiz *et al.*, 2016) for gillnets and driftnets. The integration of different satellite tracking instruments for vessels and species with geographic information systems also supports better opportunities to identify high inter-action areas. Recent studies have highlighted these sensitive convergence points with fisheries, based on bycatch reports (Lewinson *et al.*, 2014; Lezama-Ochoa *et al.*, 2019). However, the challenge is ongoing to combine tracking data of vessels in real time (i.e. Global Fishing Watch) with informa-tion on the movement of species, to implement "live" maps of potential interactions.

The progress of technologies and sensors to study the movement of migratory species generates unique opportunities to also identify critical conservation areas and habitats. A recent study combines information on movements of sharks with vessels, to evaluate the effectiveness of creating a protected marine area, to offer a safe space within the limits of some mobile species (White *et al.*, 2017). Aichi Target 11 calls on countries – including Chile, Mexico and Peru – to protect 10 percent of their coastal and marine areas by 2020.[4] The integration and analysis of different sources of informa-tion is becoming increasingly relevant to identify areas that may maximize the conservation of migratory species populations. This includes under-standing the connectivity that species establish between distant areas throughout their lives.

Spatial and temporal connectivity among areas

The identification of these two aspects is key in order to establish efficient and effective measures for the conservation of migratory species. In the case of migratory birds, it is not complicated to identify areas used during their journeys. It is easy to observe them making use of different habitats and thereby identify critical use areas to complete their seasonal movement cycles. However, this is not the case with marine species. This group presents a great challenge for researchers. For some species, information on spatial and tem-poral connectivity has only been generated in recent years. For example, a study published three years ago has monitored the movement of Whale sharks (*Rhincodon typus*) between the Galapagos Islands and Peru (Hearn *et al.*, 2016). This is one of the few studies undertaken on adult individuals. Another investigation has focused on studying the seasonal presence of juvenile whale sharks and suggests the existence of a feeding area in the north of Peru (Maguiño *et al.*, 2016).

Becoming aware of critical use areas allows for the issue of conservation strategies to be discussed. An important challenge is, for example, the conservation of land temporarily used by migratory birds and now modi-fied responding to the needs of society. A good example is based on Dynamic Conservation and can be observed in the establishment of "Pop-up-Wetlands" in rice-growing zones of the United States. This is a system of temporary rental of large extensions of rice fields in the

California Valley that allows the extension of flooding seasons for these areas – a strategy used to maintain the land – to receive thousands of birds on what was historically wetlands, now developed into agricultural zones (Reynolds *et al.*, 2017).

Conclusions and final reflections

The complexity of the life history of migratory marine species presents several challenges for their effective conservation. By occupying different and distant areas to complete a part of their lifecycle, they interact with human activities that occur in coastal-marine areas, both in shallow waters and in open seas. Faced with this reality, a synthesis of available scientific information must made as part of a critical input to the discussion, design and implementation of any public policy.

At present, there are various threats caused by human activities and the identification of future potential threats does not present a very positive outlook for migratory marine species. It is vital to identify the threats shared among countries (i.e. bycatch) and the exclusive ones for each (i.e. the extraction of eggs in Mexico) in order for appropriate regional policies.

Multilateral Environmental Agreements do exist between Mexico, Peru and Chile, to support the conservation of migratory marine species in spaces under their jurisdictions, but the effectiveness of these instruments applied to spaces and areas over which countries do not have jurisdiction (i.e. open waters) continues to be a challenge.

There are different levels of connectivity between countries as a result of the movement of migratory marine species, although this information is reduced to a few species. It is important to increase research efforts in order to identify the spaces used in their migratory routes. This would enable proposals for ad hoc protection areas and seasons that are efficient and effective to support the different life stages of migratory marine species.

In the Pacific, the threat of bycatch, marine pollution and loss of habitat puts at risk decades of efforts in critical habitats. There are examples of both successful experiences for the conservation of migratory marine species as well as unsuccessful ones. It is important to identify and analyze what works in conservation, in order to potentially transfer and integrate experiences into local and regional conservation plans under development.

Notes

1   Ministerial Resolution No. 331-2017-PRODUCE that bans the extraction of whale sharks as from July 14, 2017. Available at, www.gob.pe/institucion/produce/normas-legales/144287-331-2017-produce.
2   See, www.iacseaturtle.org/informes.htm.
3   Ibid. at 3.
4   See, Aichi Target 11, www.cbd.int/sp/targets/.

References

Acevedo, J., Aguayo-Lobo, A., Allen, J., Botero-Acosta, N., Capella, J., Castro, C., Rosa, L.D., Denkinger, J., Félix, F., Flórez-González, L. and Garita, F. (2017), Migratory Preferences of Humpback Whales between Feeding and Breeding Grounds in the Eastern South Pacific. *Marine Mammal Science*, 33(4), pp. 1035–1052.

Alfaro-Shigueto, J., Mangel, J.C., Bernedo, F., Dutton, P.H., Seminoff, J.A. and Godley, B.J. (2011), Small-Scale Fisheries of Peru: A Major Sink for Marine Turtles in the Pacific. *Journal of Applied Ecology*, 48(6), pp. 1432–1440.

Alfaro-Shigueto, J., Mangel, J.C., Dutton, P.H., Seminoff, J.A. and Godley, B.J. (2012), Trading Information for Conservation: A Novel use of Radio Broadcasting to Reduce Sea Turtle Bycatch. *Oryx*, 46(3), pp. 332–339.

Awkerman, J.A., Huyvaert, K.P., Mangel, J., Shigueto, J.A. and Anderson, D.J. (2006). Incidental and Intentional Catch Threatens Galapagos Waved Albatross. *Biological Conservation*, 133(4), pp. 483–489.

Beyer, J., Trannum, H.C., Bakke, T., Hodson, P.V. and Collier, T.K. (2016), Environmental Effects of the Deepwater Horizon Oil Spill: A Review. *Marine Pollution Bulletin*, 110(1), pp. 28–51.

Cole, M., Lindeque, P., Halsband, C. and Galloway, T.S. (2011), Microplastics as Contaminants in the Marine Environment: A Review. *Marine Pollution Bulletin*, 62(12), pp. 2588–2597.

Cruz-López, M., Eberhart-Phillips, L.J., Fernández, G., Beamonte-Barrientos, R., Székely, T., Serrano-Meneses, M.A. and Küpper, C. (2017), The Plight of a Plover: Viability of an Important Snowy Plover Population with Flexible Brood Care in Mexico. *Biological Conservation*, 209, pp. 440–448.

Derraik, J.G. (2002), The Pollution of the Marine Environment by Plastic Debris: A Review. *Marine Pollution* bulletin, 44(9), pp. 842–852.

Dittman, A. and Quinn, T., 1996. Homing in Pacific Salmon: Mechanisms and Ecological Basis. *Journal of Experimental Biology*, 199(1), pp. 83–91.

Egevang, C., Stenhouse, I.J., Phillips, R.A., Petersen, A., Fox, J.W. and Silk, J.R. (2010), Tracking of Arctic Terns (*Sterna Paradisaea*) Reveals Longest Animal Migration. *Proceedings of the National Academy of Sciences*, 107(5), pp. 2078–2081.

Frazier, J. (2014), *La Situación Regional de las Tortugas Marinas en el Pacífico Sudeste*. Comisión Permanente del Pacífico Sur.

Guayaquil, Ecuador. García-Godos, I., Waerebeek, K.V., Alfaro-Shigueto, J. and Mangel, J.C. (2013), Entanglements of Large Cetaceans in Peru: Few Records but High Risk. *Pacific Science*, 67(4), pp. 523–532.

Gilman, E., Clarke, S., Brothers, N., Alfaro-Shigueto, J., Mandelman, J., Mangel, J., Petersen, S., Piovano, S., Thomson, N., Dalzell, P. and Donoso, M. (2008), Shark Interactions in Pelagic Longline Fisheries. *Marine Policy*, 32(1), pp. 1–18.

Graham, I.M., Pirotta, E., Merchant, N.D., Farcas, A., Barton, T.R., Cheney, B., Hastie, G.D. and Thompson, P.M. (2017), Responses of Bottlenose Dolphins and Harbor Porpoises to Impact and Vibration Piling Noise During Harbor Construction. *Ecosphere*, 8(5), p.e01793.

Harrison, A.L., Costa, D.P., Winship, A.J., Benson, S.R., Bograd, S.J., Antolos, M., Carlisle, A.B., Dewar, H., Dutton, P.H., Jorgensen, S.J. and Kohin, S. (2018), The

Political Biogeography of Migratory Marine Predators. *Nature Ecology & Evolution*, 2(10), p. 1571.

Hearn, A.R., Green, J., Román, M.H., Acuña-Marrero, D., Espinoza, E. and Klimley, A.P. (2016), Adult Female Whale Sharks make Long-Distance Movements Past Darwin Island (Galapagos, Ecuador) in the Eastern Tropical Pacific. *Marine Biology*, 163(10), p. 214.

Heithaus, M.R., Wirsing, A.J., Dill, L.M. (2012), The Ecological Importance of Intact Top-Predator Populations: A Synthesis of 15 Years of Research in a Seagrass Ecosystem. *Marine and Freshwater Research*, 63: 1039–1050.

Juarez, C.A. (2004). *Determinación de Contaminantes Organoclorados en Tres Especies de Tortugas Marinas de Baja California Sur.* Trabajo de Tesis para Obtener el Grado de Maestro en Ciencias en el Uso, Manejo y Preservación de los Recursos Naturales. Centro de Investigaciones Biológicas del Noroeste, La Paz BCS, México.

Kirby, J.S., Stattersfield, A.J., Butchart, S.H., Evans, M.I., Grimmett, R.F., Jones, V.R., O'Sullivan, J., Tucker, G.M. and Newton, I. (2008), Key Conservation Issues for Migratory Land and Waterbird Species on the World's Major Flyways. *Bird Conservation International*, 18(S1), pp.S49–S73.

Lascelles, B., Notarbartolo Di Sciara, G., Agardy, T., Cuttelod, A., Eckert, S., Glowka, L., Hoyt, E., Llewellyn, F., Louzao, M., Ridoux, V. and Tetley, M.J. (2014), Migratory Marine Species: Their Status, Threats and Conservation Management Needs. *Aquatic Conservation: Marine and Freshwater Ecosystems*, 24(S2), pp. 111–127.

Lewison, R.L., Crowder, L.B., Read, A.J. and Freeman, S.A. (2004), Understanding Impacts of Fisheries Bycatch on Marine Megafauna. *Trends in Ecology & Evolution*, 19(11), pp. 598–604.

Lewison, R.L., Crowder, L.B., Wallace, B.P., Moore, J.E., Cox, T., Zydelis, R., McDonald, S., DiMatteo, A., Dunn, D.C., Kot, C.Y. and Bjorkland, R. (2014), Global Patterns of Marine Mammal, Seabird, and Sea Turtle Bycatch Reveal Taxa-Specific and Cumulative Megafauna Hotspots. *Proceedings of the National Academy of Sciences*, 111(14), pp. 5271–5276.

Lezama-Ochoa, N., Hall, M., Román, M. and Vogel, N. (2019), Spatial and Temporal Distribution of Mobulid Ray Species in the Eastern Pacific Ocean Ascertained from Observer Data from the Tropical Tuna Purse-Seine Fishery. *Environmental Biology of Fishes*, 102(1), pp. 1–17.

López-Hoffman, L., Chester, C.C., Semmens, D.J., Thogmartin, W.E., Rodríguez-McGoffin, M.S., Merideth, R. and Diffendorfer, J.E. (2017), Ecosystem Services from Transborder Migratory Species: Implications for Conservation Governance. *Annual Review of Environment and Resources*, 42, pp. 509–539.

Maguiño, R., Mendoza, A., Kelez, S., Vélez-Zuazo, X. and Ramírez-Macías, D. (2016), Unveiling a New Foraging Area for the Threatened Whale Shark. *Science Proceedings*, The 4th International Whale Shark Conference, May, p. 32.

Mangel, J.C., Alfaro-Shigueto, J., Van Waerebeek, K., Cáceres, C., Bearhop, S., Witt, M.J. and Godley, B.J. (2010), Small Cetacean Captures in Peruvian Artisanal Fisheries: High Despite Protective Legislation. *Biological Conservation*, 143(1), pp. 136–143.

Mangel, J.C., Alfaro-Shigueto, J., Witt, M.J., Hodgson, D.J. and Godley, B.J. (2013), Using Pingers to Reduce Bycatch of Small Cetaceans in Peru's Small-Scale Drift-net Fishery. *Oryx*, 47(4), pp. 595–606.

Medellín, R., Abreu-Grobois, A., Arizmendi, M.D.C., Mellink, E., Ruelas, E., Santana, E. and Urbán, J. (2009), Conservación de Especies Migratorias y Poblaciones Transfronterizas. CONABIO (ed.). *Capital Natural de México*, 2, pp. 485–490.

Meylan, A.B., Bowen, B.W. and Avise, J.C. (1990), A Genetic Test of the Natal Homing versus Social Facilitation Models for Green Turtle Migration. *Science*, 248(4956), pp. 724–727.

Miranda-Urbina, D., Thiel, M. and Luna-Jorquera, G. (2015), Litter and Seabirds Found Across a Longitudinal Gradient in the South Pacific Ocean. *Marine Pollution Bulletin*, 96(1–2), pp. 235–244.

Myers, J.P., Morrison, R.I.G., Antas, P.Z., Harrington, B.A., Lovejoy, T.E., Sallaberry, M., Senner, S.E. and Tarak, A. (1987), Conservation Strategy for Migratory Species. *American Scientist*, 75, p. 1926.

Oppel, S., Bolton, M., Carneiro, A.P., Dias, M.P., Green, J.A., Masello, J.F., Phillips, R.A., Owen, E., Quillfeldt, P., Beard, A. and Bertrand, S. (2018), Spatial Scales of Marine Conservation Management for Breeding Seabirds. *Marine Policy*, 98, pp. 37–46.

Ortiz, N., Mangel, J.C., Wang, J., Alfaro-Shigueto, J., Pingo, S., Jimenez, A., Suarez, T., Swimmer, Y., Carvalho, F. and Godley, B.J. (2016), Reducing Green Turtle Bycatch in Small-Scale Fisheries Using Illuminated Gillnets: The Cost of Saving a Sea Turtle. *Marine Ecology Progress Series*, 545, pp. 251–259.

Palacios, E., Castillo-Guerrero, J.A., Galindo-Espinosa, D., Alfaro, L., Amador, E., Fernández, G., Vargas, J. and Vega, X. (2017), Population Status of American Oystercatchers (*Haematopus palliatus frazari*) Breeding in Northwest Mexico. *Waterbirds*, 40(sp1), pp. 72–79.

Peckham, S.H., Diaz, D.M., Walli, A., Ruiz, G., Crowder, L.B. and Nichols, W.J. (2007), Small-Scale Fisheries Bycatch Jeopardizes Endangered Pacific Loggerhead Turtles. *PloS one*, 2(10), p.e1041.

Perez-Venegas, D., Pavés, H., Pulgar, J., Ahrendt, C., Seguel, M. and Galbán-Malagón, C.J. (2017), Coastal Debris Survey in a Remote Island of the Chilean Northern Patagonia. *Marine Pollution Bulletin*, 125(1–2), pp. 530–534.

Pitelka, F.A. (1979), Introduction: The Pacific Coast Shorebird Scene. *Studies in Avian Biology*, 2, pp. 1–11.

Pulido Capurro, V.M. and Bermúdez Díaz, L. (2018). Estado Actual de la Conservación de los Hábitats de los Pantanos de Villa, Lima, Perú. *Arnaldoa*, 25(2), pp. 679–702.

Reinhardt, J.F., Weaver, J., Latham, P.J., Dell'Apa, A., Serafy, J.E., Browder, J.A., Christman, M., Foster, D.G. and Blankinship, D.R. (2018), Catch Rate and At-Vessel Mortality of Circle Hooks versus J-hooks in Pelagic Longline Fisheries: A Global Meta-analysis. *Fish and Fisheries*, 19(3), pp. 413–430.

Reynolds, H.L. and Clay, K. (2011), Migratory Species and Ecological Processes. *Environmental Law*, 41, p. 371.

Reynolds, M.D., Sullivan, B.L., Hallstein, E., Matsumoto, S., Kelling, S., Merrifield, M., Fink, D., … and Morrison, S.A. (2017), Dynamic Conservation for Migratory Species. *Science Advances*, 3(8), p.e1700707.

Ribeiro, S., Viddi, F.A. and Freitas, T.R. (2005), Behavioural Responses of Chilean Dolphins (*Cephalorhynchus eutropia*) to Boats in Yaldad Bay, Southern Chile. *Aquatic Mammals*, 31(2), p. 234.

Robbins, C. S., Sauer, J. R., Greenberg, R. S. and Droege, S. (1989), Population Declines in North American Birds that Migrate to the Neotropics. *Proceedings of the National Academy of Sciences*, 86: 7658–7662.

Runge, C.A., Watson, J.E., Butchart, S.H., Hanson, J.O., Possingham, H.P. and Fuller, R.A. (2015), Protected Areas and Global Conservation of Migratory Birds. *Science*, 350(6265), pp. 1255–1258.

Sauer, J. R., Hines, J. E. and Fallon, J. (2005), *The North American Breeding Bird Survey, Results and Analysis* 1966–2005. Laurel, MD: USGS Patuxent Wildlife Research Centre.

Suazo, C.G., Cabezas, L.A., Moreno, C.A., Arata, J.A., Luna-Jorquera, G., Simeone, A., Adasme, L., Azócar, J., García, M., Yates, O. and Robertson, G. (2014), Seabird Bycatch in Chile: A Synthesis of its Impacts and a Review of Strategies to Contribute to the Reduction of a Global Phenomenon. *Pacific Seabirds*, 41, pp. 1–12.

Sutherland, W.J., Alves, J.A., Amano, T., Chang, C.H., Davidson, N.C., Max Finlayson, C., Gill, J.A., Gill Jr, R.E., González, P.M., Gunnarsson, T.G. and Kleijn, D. (2012), A Horizon Scanning Assessment of Current and Potential Future Threats to Migratory Shorebirds. *Ibis*, 154(4), pp. 663–679.

Thiel, M., Bravo, M., Hinojosa, I.A., Luna, G., Miranda, L., Núñez, P., Pacheco, A.S. and Vásquez, N. (2011), Anthropogenic Litter in the SE Pacific: An Overview of the Problem and Possible Solutions. *Revista de Gestão Costeira Integrada-Journal of Integrated Coastal Zone Management*, 11(1), pp. 115–134.

Valenzuela, L.O., Sironi, M., Rowntree, V.J. and Seger, J. (2009), Isotopic and Genetic Evidence for Culturally Inherited Site Fidelity to Feeding Grounds in Southern Right Whales (*Eubalaena australis*). *Molecular Ecology*, 18(5), pp. 782–791.

Velez-Zuazo, X., Ramos, W.D., Diez, C.E., Abreu-Grobois, A. and McMillan, W.O. (2008), Dispersal, Recruitment and Migratory Behaviour in a Hawksbill Sea Turtle Aggregation. *Molecular Ecology*, 17(3), pp. 839–853.

Velez-Zuazo, X., Mangel, J.C., Seminoff, J.A., Wallace, B.P. and Alfaro-Shigueto, J. (2017), Filling the Gaps in Sea Turtle Research and Conservation in the Region Where it Began: Latin America. *Latin American Journal of Aquatic Research*, 45(3), pp. 501–505.

White, T.D., Carlisle, A.B., Kroodsma, D.A., Block, B.A., Casagrandi, R., De Leo, G.A., Gatto, M., Micheli, F. and McCauley, D.J. (2017), Assessing the Effectiveness of a Large Marine Protected Area for Reef Shark Conservation. *Biological Conservation*, 207, pp. 64–71.

Žydelis, R., Small, C. and French, G., 2013. The Incidental Catch of Seabirds in Gillnet Fisheries: A Global Review. *Biological Conservation*, 162, pp. 76–88.

10 Prevention of marine pollution from litter in Peru

Irene Hofmeijer

Introduction

Oceans regulate climate, provide food resources and contribute significantly to marine-related industries which in some cases are key pillars in national economies. However, this same source of life and resources has also been treated as an open litter and waste dump throughout human history.[1] The exponential growth of the world population and industrial development during the last two centuries has resulted in litter reaching levels which oceans cannot assimilate and process.

But things are changing. The first wave of change appeared in the 1970s, when a series of international conventions banned the practice of "ocean dumping" – dumping litter and waste into the sea.[2] Although these agreements seek to regulate the transport and disposal of litter directly into the sea they do not regulate disposal into aquatic and terrestrial ecosystems (e.g. rivers, lakes and wetlands) that eventually flow into the sea. For a long time, in part due to the idea of the vastness of oceans and abundance of sea life, common materials from municipal litter, waste and garbage such as plastics, glass and metals were not perceived as a threat to marine life (Laist, 1987).

A second wave of change started to take shape in the late 1980s, when scientists began to demonstrate the negative impacts of litter on marine life (Laist, 1987). In 1995, the Washington Declaration on the Protection of the Marine Environment from Land Based Activities acknowledged the effects of waste, garbage, debris and sewage, among others, as a source of ocean pollution. The concept of "marine waste or litter" can be broadly defined as any persistent solid material that is manufactured or processed, disposed into or abandoned near coastal and marine environments (CPPS, 2007).[3]

This general trend of wakening up to marine pollution has been consistent throughout the new century, strongly influenced by advocacy and scientific research that led to the discovery of the "Great Pacific Garbage Patch" by Captain Charles Moore in 1997, in the North Pacific (Moore and Phillips, 2011). The visual impact of this discovery coupled with convincing scientific evidence, positioned marine litter on the global agenda.

The third and current wave which can be classified as the "taking action phase", started around 2010 when the United Nationals Environment (former United Nations Environment Program), declared plastic waste in the oceans a challenging and emerging issue for the global environment.[4] UN Environment emphasized plastic given that it represents 80 to 90 percent of marine litter (UNEP, 2011). Since then, research has proliferated demonstrating the impacts of plastic pollution on marine ecosystems and human health. This has to be added to material damages caused by unregulated industrial fishing, shipping fleets and tourism. To prevent marine pollution due to litter, a wide range of initiatives led by the public and private sectors and civil society have also multiplied. This has resulted in major advances in *prevention*,

Annex 10.1 Honolulu Strategy

Global framework for prevention and management of marine debris

Goal A: Reduced amount and impact of land-based sources of marine debris introduced into the sea

Strategy A1. Conduct education and outreach on marine debris impacts and the need for improved solid waste management.

Strategy A2. Employ market-based instruments to support solid waste management, in particular waste minimization.

Strategy A3. Employ infrastructure and implement best practices for improving stormwater management and reducing discharge of solid waste into waterways.

Strategy A4. Develop, strengthen, and enact legislation and policies to support solid waste minimization and management.

Strategy A5. Improve the regulatory framework regarding stormwater, sewage systems, and debris in tributary waterways.

Strategy A6. Build capacity to monitor and enforce compliance with regulations and permit conditions regarding litter, dumping, solid waste management, stormwater, and surface runoff.

Strategy A7. Conduct regular cleanup efforts on coastal lands, in watersheds, and in waterways – especially at hot spots of marine debris accumulation.

Goal B: Reduced amount and impact of sea-based sources of marine debris, including solid waste; lost cargo; abandoned, lost, or otherwise discarded fishing gear (ALDFG); and abandoned vessels, introduced into the sea.

Strategy B1. Conduct ocean-user education and outreach on marine debris impacts, prevention, and management.

Strategy B2. Develop and strengthen implementation of waste minimization and proper waste storage at sea, and of disposal at port reception facilities, in order to minimize incidents of ocean dumping.

Strategy B3. Develop and strengthen implementation of industry best management practices (BMP) designed to minimize abandonment of vessels and accidental loss of cargo, solid waste, and gear at sea.

Strategy B4. Develop and promote use of fishing gear modifications or alternative technologies to reduce the loss of fishing gear and/or its impacts as ALDFG.

Strategy B5. Develop and strengthen implementation of legislation and policies to prevent and manage marine debris from at-sea sources, and implement requirement of MARPOL Annex V and other relevant international instruments and agreements.

Strategy B6. Build capacity to monitor and enforce (1) national and local legislation, and (2) compliance with requirements of MARPOL Annex V and other relevant international instruments and agreements.

Goal C: Reduced amount and impact of accumulated marine debris on shorelines, in benthic habitats, and in pelagic waters

Strategy C1. Conduct education and outreach on marine debris impacts and removal.

Strategy C2. Develop and promote use of technologies and methods to effectively locate and remove marine debris accumulations.

Strategy C3. Build capacity to co-manage marine debris removal response.

Strategy C4. Develop or strengthen implementation of incentives for removal of ALDFG and other large accumulations of marine debris encountered at sea.

Strategy C5. Establish appropriate regional, national, and local mechanisms to facilitate removal of marine debris.

Strategy C6. Remove marine debris from shorelines, benthic habitats, and pelagic water.

---

such as the Honolulu Strategy – A Global Framework for the Prevention and Management or Marine Debris (see Annex 10.1) (UNEP and NOAA, 2011); action plans against marine debris from the UN Environment Regional Seas Program (CPPS, 2007)[5]; and national legislation that regulates the use of disposable plastics (UNEP, 2018).

Latin America has been slow to adopt regulations for disposable plastics, the main source of terrestrial and marine litter (Xanthos and Walker, 2017). In June 2018, Chile became the first country in the region to adopt a law on the use of plastic bags: Peru followed this example, and in December 2918, enacted a law to regulate single-use plastics. Other examples include local-level regulations banning plastic bags such as in the case of Querétaro and Tijuana, Mexico, in 2018.

This chapter looks at marine litter prevention in Peru, under the framework provided by the Honolulu Strategy. The chapter first determines the source of marine litter, then describes some national initiatives in Peru to prevent marine litter from land-based sources, and finally analyses what is the state of marine litter in Peru, including some of the challenges and opportunities the country and by extension Latin America may have in confronting this problem.

Sources of marine litter

Marine litter is a global phenomenon with long-lasting environmental, social and economic implications. It is also a multidimensional problem of a complex cultural nature (Coe and Rogers, 2012). To prevent marine pollution from litter it is necessary to understand where it comes from.

Land-based or marine-based litter in its origin, consists of any solid material that is processed as a result of anthropogenic activities and deliberately discharged into seas, rivers or coasts; discharged indirectly into the ocean through rivers, residual waters, storm waters, waves or winds; and dumped accidentally due to bad weather or abandoned deliberately by people on the beaches or coasts (UNEP, 2005).

Sea-based sources include ships (merchant, ferries and cruisers), fishing vessels, military and research fleets, recreational boats (yachts, sailboats, etc.), oil and gas platforms, and aquaculture infrastructure. Land-based sources include dumps, municipal landfills, river transport of waste, discharge of wastewater and untreated storm waters, leaks from industrial installations, land transport, illegal landfills, coastal tourism and micro-plastics not captured by water treatment systems for cosmetic products and washing of synthetic fibers (polyester and acrylic) (UNEP, 2005; Niaounakis, 2017). It is estimated that 20 percent of marine litter originates from sea-based sources and 80 percent from land-based sources (Niaounakis, 2017).

There is less pollution from sea-based sources due to international regulations that ban the disposal of litter and waste at sea and voluntary efforts by the marine industry. Nevertheless, difficulties persist in the implementation of these regulations. Furthermore, the increase of fishing and aquaculture activities is presenting new sources of pollution due to the loss of materials in the sea, many times highly durable plastics (Niaounakis, 2017). The greatest challenge to avoid continuous deterioration of marine ecosystems due to litter is preventive action on land.

Prevention of marine litter from land-based sources in Peru

Ultimately, the prevention of marine pollution from litter and waste originated from land-based sources requires good management practices and minimizing the production of solid waste, particularly plastics (Jambeck et al., 2015). It is estimated that between 4.8 and 12.7 million tons of plastics entered the oceans in 2010 due to global mismanagement of litter. With the projections of population growth and without significant improvements in solid waste management systems, numbers could multiply tenfold by 2025 (Jambeck et al., 2015).

Chile, Mexico and Peru are emerging economies where important changes in consumption habits are taking place. The economic growth of Peru is one of the highest in South America, reaching an average of 6.3 percent from 2005 to 2014 (Rossini and Santos, 2015). This growth has resulted in a middle class that now represents 40 percent to 50 percent of the population. They have greater economic power and, thereby, have increased consumption (Jaramillo and Zambrano, 2013; Zapatel, 2012). Globally, the North American and European "dream" influences consumption habits of the emerging middle class: an unsustainable high consumer lifestyle that generates large volumes of waste and litter (Akenji and Chen, 2016).

To avoid ocean "plastification", a multi-sectorial effort is required. This means various things: raising awareness among citizens about sustainable consumption; extending pollution liability to waste and litter producers; and developing adequate infrastructure and policy frameworks to minimize, manage and dispose of litter. The first goal of the Honolulu Strategy proposes seven lines of action to reduce the amount and impact of land-based sources of marine debris (litter) introduced into the sea. Following are the lines of action and activities currently under way.

A1 Conduct education and awareness campaigns on marine debris and the need for improvement in solid waste management

Although environmental awareness among the general public has increased in Latin America, the habit of throwing away garbage and waste persists (Thiel *et al.*, 2011). At present, Peru has a series of initiatives from different sectors underway to encourage a responsible consumer culture and care for the sea and ocean (see Annex 10.2). These campaigns have strengthened during the last decade, thanks to the efforts and leadership of civil society organizations.

In the late 1990s, citizens first mobilized in Peru to address marine litter from land-based sources. They were led by civil society through groups such as Vida (Life) and Ecoplayas (Ecobeaches). In 2000, the NGO Ciudad

Annex 10.2 Leading Peruvian organizations dedicated to promoting responsible consumption and caring for the sea

Organization	Type	Year of constitution	Web page
Vida – Instituto para la Protección del Medio Ambiente	Civil Association	1990	www.vida.org.pe
Ecoplayas	Civil Association	1997	www.ecoplayasperu.wixsite. com/rediberoamericana
Ciudad Saludable	NGO	2002	www.ciudadsaludable.org
Planeta Océano	NGO	2007	www.planetaoceano.org
Organization	Type	Year of constitution	Web page
Life Out Of Plastic – L.O.O.P.	Social Company	2011	www.loop.pe
Conservamos por Naturaleza	NGO	2011	www.conservamospornaturaleza. org
HAZla Por tu Playa	Campaign	2012	www.hazla.pe
B-Green	Consultant	2012	www.b-green.pe
Oceana Peru	NGO	2015	www.peru.oceana.org

Note
The list is limited to leading organizations that over the past three years have directly implemented campaigns related to land-based marine litter. It should be noted that in recent years a number of initiatives have emerged, noticeably on social networks that promote responsible consumption and beach cleaning.

Saludable (Sustainable City) started substantial work in the area of solid waste management, starting with the formalization of recyclers, awareness-raising campaigns and direct municipal interventions. Important as they were, these initiatives were very localized in their influence and impacts.

During the past decade, young entrepreneurs and activists have had growing influence through social networks promoting a culture of protection for the sea and oceans and moving to plastic free lifestyles. We Conserve for Nature, L.O.O.P. and its joint Do It for your Beach campaign, have had a strong impact. Thousands of people have been mobilized throughout the country, raising awareness and triggering action from local communities to act upon marine pollution produced by litter (Naveda *et al.*, 2019).

These environmental awareness campaigns, jump-started by civil society, have also triggered positive action by the state. In 2013, the Multi-Sectorial Commission for the Environmental Management of the Coastal and Marine Area was created and is led by the Ministry of the Environment. In 2015, the Multi-Sectorial Commission formed a technical working group specialized in environmental education, communication and social empowerment, with an area of work focused on marine litter. Additionally, in 2018 a technical working group was created specifically dedicated to marine litter and waste.

The National Plan for Environmental Education (2016) of the Ministry of Education is an important institutional structure. The National Plan divides environmental education into four major areas: climate change, eco-efficiency, health and risk management. Under eco-efficiency, sustainable consumption and solid waste management is a key priority. As a result, the Ministry of Education has mandated that all education institutions should undertake activities and have campaigns in place regarding marine litter and solid waste management.

A2 Adopt market-based instruments to support solid waste management, particularly waste reduction

Market-based instruments include policies to stimulate sustainable purchases, taxes on certain types of materials and incentives for development of products that generate less solid waste and litter. Various laws and regulations in Peru include market-based instruments to support the management of solid waste and waste reduction (see Annex 10.3). Notably, the Law that Regulates Single-Use Plastics and Disposable Containers enacted in December 2018 has two important market-based instruments: a minimum percentage of recycled plastic in containers (15 percent) and a tax on plastic bags.

However, economic incentives to contribute to the development of new production models are still lacking. The online library of Innovate Perú of the Ministry of Production only identifies one innovation (among 180 entries) that contributes to waste and litter reduction (Innóvate Perú). At present, public agencies do not have programs to encourage innovations in a circular economy or improve solid waste management. Nor are there market-based tools to promote the improvement of waste management, such

Annex 10.3 Examples of relevant Peruvian laws for the prevention of marine waste and litter from land-based sources

Legal norm	Title	Objective	Type of prevention
Supreme Decree 009-2009-MINAM	Eco-efficiency measures for the public sector.	Implement solid waste segregation and recyclable systems in public sector institutions.	Improvements in the integrated management of solid waste and litter.
Supreme Decree 004-2011-MINAM	Gradual application of the percentages of recycled material (plastic, paper and carton) the public sector should use and buy.	Generate incentives for the use of recyclable materials in the public sector.	Public procurement encouraging the purchase of products derived from a circular economy.
Law 30,884	Law that regulates single-use plastics and disposable containers or packages.	Minimize the use of disposable plastics.	Ban disposable materials. Encourage a circular economy.
Supreme Decree 014-2017-MINAM	Law on solid waste integrated management.	Promote adequate provisions for solid waste at different government levels.	Improve solid waste integrated management. Encourage a circular economy.

as a litter collection taxes or deposit programs for certain high-value materials like aluminum, PET plastics or electronic waste.

A3 Adopt infrastructure and implement best practices to improve rainwater management and reduce the discharge of solid waste into waterways

Approximately 58 percent of all litter and solid waste in Peru is disposed of in open-air landfills, often near waterways. Less than 4 percent is recycled (Bird and Weaver, 2018; MINAM, 2017). Peru has a deficit in necessary infrastructure to manage 7.5 million tons of municipal solid waste generated each year (MINAM, 2017). The Ministry for the Environment estimates that proper management would require 190 properly managed landfills for the final disposal of solid waste and litter (MINAM, 2017). This law reduces the approval process of technical documents for building landfills from five years to one year, facilitating the construction of new infrastructure.[6]

Another problem apart from limited infrastructure is the absence of an adequate system for the recovery of reusable waste and litter. Recycling programs are limited to municipal programs that do not provide an integrated and comprehensive service in all districts. Although there are social responsibility initiatives led by the private sector to establish pickup points for the recovery of recyclables, there is no public data to evaluate their efficiency and contribution to the improvement of solid waste and litter management.

Inexistence of infrastructure is especially problematic when natural disasters, such as "*huaycos*" (landslides), overload and collapse waterways and result in large volumes of waste and litter entering the sea. Existence of projects to manage large amounts of storm waters is unknown and information is not available.

A4 Develop, strengthen and enact legislation and policies to support the minimization and management of solid waste and litter

Peru has a set of laws (see Annex 10.3) that support minimization and management of solid waste. There has been comprehensive legislation on solid waste and litter management since 2000. However, advances have been limited, largely due to uncoordinated actions and implementation at the local municipal level. To a great extent, this is also a result of neither financial resources nor technical capacities being available to implement waste management projects at the local municipal level (Bird and Weaver, 2018). The new Law on Integrated Solid Waste Management includes important improvements.[7] The law stands on three pillars: reducing the generation of waste, efficiency in the use of materials and encouraging a circular economy that reuses waste and litter through recycling chains. Hopefully this will unblock processes and investment to stimulate a recycling and reuse industry (MINAM, 2017).[8]

Additionally, the Law that Regulates Single-Use Plastics and Disposable Containers is also an important milestone to disincentivize and minimize the generation of solid waste and litter, particularly in the form of plastics.[9] The law establishes a prohibition to use single-use plastics in natural protected areas, beaches, museums, World Heritage Sites and in state buildings. It extends this prohibition to dispensing single-use plastics in supermarkets, auto services and businesses in general.

A5 Improve the legal framework on rain waters, wastewaters and debris in tributaries and waterways

At present, only 15 percent of wastewater is treated in Peru (El Comercio, 2017). Together with a non-existent integrated solid waste management system, this means that rivers full of waste and litter flow into the sea. The Framework Law for the Management and Provision of Sanitation Services was enacted in 2016 and provides for the setting up of a fund managed by the Ministry of Housing, Construction and Sanitation for the construction of wastewater treatment plants.[10] Additionally, Water Resources Law was modified in order for wastewater treatment plants to progressively adapt to the environmental management instruments, with a timeframe of nine years for sanitation service providers to comply with quality discharge standards and regulate the discharge or overflow of wastewater resulting from sanitation services.[11]

A6 Build capacity to monitor and enforce regulations and conditions on permits to leave the garbage on the street, to throw away large volumes of garbage, to manage wastewater, rain water and surface runoff

Although a legal framework in Peru for wastewater management has been in place since 2000, when the first General Law on Solid Waste and Law on Integrated Solid Waste Management were enacted, implementation and enforcement have been also limited. This was accompanied by the National Plan for Solid Waste Management in 2005. The application of its guidelines and supervision of the law have also been limited. By 2015, only 176 of 1,855 municipalities in the country (9.5 percent) had implemented programs of waste separation at the source and selective collection of solid waste (MINAM, 2017).

In 2008, the Solid Waste Management Information System was implemented. The System registers data regarding solid waste collection by municipalities. However, the data generated has considerable gaps. Approximately 90 percent of recyclables are collected and accumulated by informal actors who operate outside the regulatory framework and public system and so data is not reflected in the System. The Environmental Assessment and Control Agency of the Ministry of the Environment was created in 2008,[12] and is in charge of environmental control and seeking a balance between private investment in economic activities and environmental protection. The Control Agency supervises activities and problems related to waste disposal at the provincial municipal level in particular.

A7 Make regular efforts to clean the coast, watersheds and waterways – mainly at critical spots where marine debris is accumulated

Efforts to clean the coastline, watersheds and waterways in Peru are mainly triggered and supported through civil society action. For example, from 2013 to 2018 the Do it for your Beach campaign has mobilized 18,100 volunteers who conducted 943 clean-ups and collected 154 tons of litter (Naveda *et al.*, 2019). However, the main purpose of these clean-up campaigns is to raise awareness regarding marine waste and litter and its impacts. These are not regular, nor are they institutional clean-up efforts led by responsible and competent state agencies and municipalities. Although they often participate in these campaigns they are not part of a rolling and planned agenda to prevent marine pollution from waste and litter.

Present situation of marine litter in Peru

Even though the situation is far from optimum, the previous section has shown how certain advances have been made in Peru with regards to the first goal of the Honolulu Strategy and marine pollution prevention from land-based waste and litter. A broad set of legal instruments have been enacted which seek to minimize the production of waste and litter, improve integrated solid waste management and encourage a citizens' culture to care for and appreciate the environment. Civil society movements and campaigns have served to create awareness and encourage changes in peoples' behavior with respect to waste, litter and the ocean. This is the positive side.

Nevertheless, a limiting factor in strengthening good governance, management of waste and litter and improving social action in Peru, is limited production and dissemination of sound data and information to improve decision making and better understand the impact of waste and litter on marine and coastal environments. Information and reports are scarce and limited in their scope. Apart from studies which quantify how much waste is generated in different sectors, there is no breakdown and analysis regarding the specific impacts of land-based waste and litter on coasts and marine ecosystems. For instance, little is known about the impacts of tourism, industrial fisheries and marine-related industries in terms of the costs they externalize in terms of waste and litter disposed in the sea or which reaches the sea by different means. The environmental, social and economic costs of this impact are, for the most part, unknown.

Limited data and information affect the production of good laws and regulations. Likewise, good and sound general public policies cannot be discussed and passed in absence of data and information. One area of increasing interest for advocates of clean seas and healthy environments are micro-plastics from cosmetics which are marketed in Peru, but which have been banned in various countries around the world. Good data and information about their use and impacts in Peru specifically is simply unavailable.

Opportunities and challenges

This chapter, which has focused on Peru, is probably applicable to other countries in the region. One interesting fact is that, overall, Latin America is not – in relative terms – a large contributor to marine litter. Therefore, it has an opportunity to position itself as a global leader and may be in a position to generate zero litter which affects costal and marine environments or simply to close the flow of plastics and litter which reach the sea and coasts. Whilst ambitious, the opportunity and present context offer a major possibility to achieve this.

Over the past decade or so, Peru has made considerable progress in terms of its regulatory framework, which clearly targets the production and flow of land-based waste and litter into the sea. The challenge, however, centers on how to improve implementation and enforcement regarding minimizing and managing waste, how to create a circular economy where waste is minimized and continuously recycled and how to produce environmentally responsible citizens in a developing country with pressing needs in many other sectors.

To overcome limitations of solid waste management systems, regional and municipal governments need to strengthen their monitoring and technical capacities. Improving the quality of data and information regarding solid waste could allow market forces to create incentives and attract new investments in the recycling value chain. Strengthening systems for the segregation and collection of recyclables presents opportunities for the development of public-private alliances that could, at the same time, strengthen the recyclable market.

The creation of a high-quality recyclable market opens the doors for the development of new industries as part of the circular economy. Peru, and other countries in Latin America in general, have an opportunity to skip the old model of giant landfills for solid waste management. They can create intelligent systems for integrated solid waste management that could reuse materials efficiently.

The greatest challenge, however, which triggers all other processes, is the development of a citizen environmental culture. With a growing middle class together with a capacity for consumption, the challenge is to encourage sustainable consumption livelihoods. Although the recycling industry is key, minimizing production of waste is even more efficient and useful. Peru faces an uphill battle in culture change, given people have become accustomed to throwing waste and litter even onto the street.

Successful cases in European and African countries have shown that a stronger measures for behavioral changes is tax imposition on disposable plastics. In Ireland, for example, a 90 percent reduction of plastic bag use was achieved, after the introduction of a tax on plastic bags. A similar measure was taken in Rwanda, the first African country to ban plastic bags. We expect to see similar results in Peru when the ban established in the new law on single-use plastics enters into force in 36 months.

Final reflections

One cannot remain passive in the face of the impact of plastic pollution and marine waste and litter. During the last decade the world has witnessed a transformation of the marine ecosystem, from a blue world full of riches to the largest landfill of plastics. Responsibility for actions to prevent marine litter lies at all levels: personal, national, regional and international. Peru has the necessary pieces of the puzzle to prevent waste and litter from land-based sources entering the sea. To accelerate the process of change, the quality of data collected needs to improve in order to measure the efficiency of the impact of the work being undertaken at present. Urgent analysis needs to be made of the impact of waste and litter on the Peruvian marine ecosystem and the national economy. Furthermore, the fight against marine pollution from land-based sources is closely connected to generating a behavioral change in citizens. The current situation offers important opportunities to promote efficient industries at different stages of a circular economy.

Notes

1 Often, the concepts of trash, debris, litter, waste and garbage are used as synonyms and interchangeably in the literature and legal instruments. The chapter prefers the use of "litter" and "waste" as more general and encompassing.
2 The most relevant international conventions and agreements for marine litter include: the United Nations Convention on the Law of the Sea (UNCLOS); Annex V of the International Convention for the Prevention of Pollution from Ships (MARPOL); the London Convention on the Prevention of Marine Pollution by Dumping of

Wastes and other Matter; the Stockholm Convention on Persistent Organic Pollutants; the Rotterdam Convention on the Prior Informed Consent Procedure for Certain Hazardous Chemicals and Pesticides in International Trade; the Basel Convention on the Control of Transboundary Movements of Hazardous Wastes.

3 The Washington Declaration on the Protection of the Marine Environment (1995) where over 108 governments committed to the protection and preservation of the marine environment from the impacts of land-based activities.

4 See, www.unenvironment.org/es.

5 The UN Environment Regional Seas Program is responsible for the implementation of UN Environment marine policies to prevent the degradation of marine eco-systems and coastal areas. It seeks to unite neighboring countries to develop coordinated actions in shared marine areas. In South America, Chile, Peru, Ecuador and Colombia are grouped under the Permanent Commission for the South Pacific (CPPS). In 2007, the CPPS produced i) a report analyzing the production of solid waste and litter from land-based and marine sources which could become marine litter for the region and ii) a regional program for integrated management of marine litter in the South Pacific (CPPS, 2007).

6 MINAM (2018b), Nueva Ley de Residuos Sólidos, accessed 15 December 2018. Availableat,www.minam.gob.pe/gestion-de-residuos-solidos/nueva-ley-de-residuos-solidos/.

7 Legislative Decree 1278 of 23 December 2016, available at: https://sinia.minam. gob.pe/normas/ley-gestion-integral-residuos-solidos.

8 Ibid, at 6.

9 Law 30.884 of 19 December 2018, available at, https://sinia.minam.gob.pe/ normas/ley-que-regula-plastico-un-solo-uso-recipientes-envases-descartables.

10 Legislative Decree 1280 of 28 December 2016, available at, https://busque-das.elperuano.pe/normaslegales/decreto-legislativo-que-aprueba-la-ley-marco-de-la-gestion-y-decreto-legislativo-n-1280–1468461–1/.

11 Law 29.330 of 30 March 2009, available at, www.ana.gob.pe/sites/default/files/ publication/files/ley_29338_0.pdf.

12 The Environmental Assessment and Control Agency was created by Legislative Decree 1,013 on 30 May 2008, see, www.oefa.gob.pe/normas-de-creacion.

References

Akenji, L., and Chen, H. (2016), A Framework for Shaping Sustainable Lifestyles: Deter-minants and Strategies. UN Environment. Available at, www.oneplanetnetwork.org/ sites/default/files/a_framework_for_shaping_sustainable_lifestyles_determinants_and_ strategies_0.pdf.

Bird, T., and Weaver, S. (2018), Climate Action in Peru: Nordic Support for Waste Sector Management Yields Results. In: *ANP*. Copenhagen: Nordisk Ministerråd.

Coe, J., and Rogers, D. (Eds.) (1997), *Marine Debris: Sources, Impacts, and Solutions* (Springer Series in Environmental Management. New York: Springer Verlag.

CPPS. (2007), *Basura Marina en el Pacífico Sudeste: una Revisión del Problema*. Guayaquil, Ecuador: Comisión Permanente del Pacífico Sur.

El Comercio (2017), ¿Existen Sistemas para Tratar Aguas Residuales en el Perú? Available at, https://elcomercio.pe/economia/peru/existen-sistemas-tratar-aguas-residuales-peru-noticia-455379.

Innóvate Perú. 'Biblioteca Digital'. Accessed 15 December 2018. www.innovateperu. gob.pe/quienes-somos/biblioteca-digital.

Jambeck, J., Geyer, R., Wilcox, C., Siegler, T.S., Perryman, M., Andrady, A., Narayan, R., and Lavender Law, K. (2015), Plastic Waste Inputs from Land into the Ocean, *Science*, 347: 768–71.

Jaramillo, F., and Zambrano, O. (2013), La Clase Media en Perú: Cuantificación y Evolución Reciente. In: *IDB Technical Note*. Interamerican Development Bank. Available at, file:///C:/Users/MANOLO/Downloads/La-clase-media-en-Per%C3%BA-Cuantificaci%C3%B3n-y-evoluci%C3%B3n-reciente.pdf.

Laist, D. (1987), Overview of the Biological Effects of Lost and Discarded Plastic Debris in the Marine Environment, *Marine Pollution Bulletin*, 18: 319–26.

MINAM (2017), *Plan Nacional de Gestión Integral de los Residuos Sólidos*. Ministerio del Ambiente, 85. Lima: Ministerio del Ambiente – MINAM.

MINAM (2018a), 'Listado de Rellenos Sanitarios'. Accessed 1 February 2019. www.minam.gob.pe/wp-content/uploads/2018/12/listado-de-rellenos-sanitarios-_4-12-2018_v2.pdf.

MINAM (2018b), Nueva Ley de Residuos Sólidos, accessed 15 December 2018. Available at, www.minam.gob.pe/gestion-de-residuos-solidos/nueva-ley-de-residuos-solidos/.

Moore, C., and Phillips, C. (2011), *Plastic Ocean: How a Sea Captain's Chance Discovery Launched a Determined Quest to Save the Oceans*. Avery: New York.

Naveda, M., Monteferri, B., Balducci, N., Lo, J., Loli, P., Hofmeijer, I., Ruiz, A., Roldán, P., Wust, W., and Contreras, F. (2019), *HAZla por tu Playa 2013–2018*. Available at, http://hazla.pe/about-us-1.

Niaounakis, M. (2017), *Management of Marine Plastic Debris. Prevention, Recycling and Waste Management*. Oxford, UK; Cambridge, MA: Elsevier, William Andrew.

Rossini, R., and Santos, A. (2015), Peru's Recent Economic History: From Stagnation, Disarray, and Mismanagement to Growth, Stability, and Quality Policies. In: Werner, A., and Santos, A. (eds.), *PERU: Staying the Course of Economic Success* (International Monetary Fund: Washington, DC).

Thiel, M., Bravo, M., Hinojosa, I., Luna, G., Miranda, L., Núñez, P., Pacheco, A., and Vásquez, M. (2011), Anthropogenic Litter in the SE Pacific: An Overview of the Problem and Possible Solutions, *Revista de Gestão Costeira Integrada-Journal of Integrated Coastal Zone Management*, 11 (1) 115-134.

UNEP (2005), *Marine Litter, An Analytical Overview*. Available at http://wedocs.unep.org/bitstream/handle/20.500.11822/8348/-arine%20Litter%2c%20an%20analytical%20overview-20053634.pdf?sequence=3&isAllowed=y

UNEP (2011), *UNEP YEAR BOOK 2011: Emerging Issues in our Global Environment*. Nairobi, Kenya.

UN Environment (2018), *Single-Use Plastics: A Roadmap for Sustainability*. Nairobi, Kenya.

UNEP and NOAA. (2011), The Honolulu Strategy: A Global Framework for Prevention and Management of Marine Debris. Available at, https://repository.library.noaa.gov/view/noaa/10789.

Xanthos, D., and Walker, T.W. (2017), International Policies to Reduce Plastic Marine Pollution from Single-Use Plastics (Plastic Bags and Microbeads): A Review, *Marine Pollution Bulletin*, 118: 17–26.

Zapatel, A. (2012), Media y con Clase. In: *Semana Económica*. Lima, Peru.

11 Collective action spaces and transformations in the governance of fisheries resources

Towards democratic and deliberative management

Rodrigo A. Estévez and Stefan Gelcich

Introduction

Sustainable management of common pool resources is one of the most significant social and environmental problems of the twenty-first century, particularly in the case of vulnerable communities. For instance, fish and other marine resources play a key role in the diet and livelihood of thousands of people all over the world (Bene *et al.*, 2005). However, due to the overexploitation of common pool resources, the growing anthropogenic pressures on the environment and structural inequalities regarding access, commercialization and use of these resources, there is a widespread recognition about the need to transform governance models where participation, transparency and equity are central dimensions (Ostrom, 2005).

Common pool resources are identified based on two specific features that distinguish them from "common goods" and "open resources". Common pool resources are extractable, in other words, the extraction of a unit implies there would be fewer available units for the rest of the users. However, common goods do not necessarily have this characteristic. For example, a farmer using solar energy for his crops does not prevent or reduce the possibility for another farmer to also use solar energy for his crops. In this context, solar energy is a common good for farmers. Second, it is hard to exclude third parties from common resources. In other words, to restrict the use and extraction of resources for external users without a right assigned or granted is relatively difficult (Schlager and Ostrom, 1992). The formal right of a group of individuals to extract a common pool resource essentially limits the right for the rest of the individuals to extract the same resource (Ostrom, 2011; Baur and Binder, 2013). The access to and use of restrictions differentiate common pool resources from open resources (Ostrom, 1992).

The "tragedy of the commons" originally presented by Hardin (1968), has turned into a metaphor for the overexploitation of common pool resources. The tragedy predicts that individuals will tend to extract these resources until their depletion, with selfish conducts prevailing over the common good (Ostrom, 2011; Hardin, 1968). In the absence of a natural predisposition towards cooperation, a solution to the tragedy of the commons for some would be to increase control from a central state, and for others, the deregulation of markets, promoting privatization of common pool resources (Hardin, 1968).

However, based on the theoretical framework of Institutional Analysis and Development, Elinor Ostrom and her colleagues demonstrate, with historiographical soundness, the development of self-regulated institutional forms to administrate common resources (Ostrom, 2011; Bromley, 1992). In other words, the tragedy highlighted by Hardin, which could only be reversed through action from a centralized state or a market that promotes privatization and accumulation of capital, can also be addressed based on the capacity of self-organizing, internal regulation, self-control and self-monitoring mechanisms of users that access common pool resources.

The development of self-organizing systems for the administration of common pool resources constitutes what could be conceived as a polycentric system. This implies the creation of several relatively autonomous decision-making units with a capacity to establish regulations, self-monitor and self-control (Ostrom, 2010). Polycentric systems allow for the creation of rules and procedures based on local knowledge of associated communities (Ostrom, 1992). The polycentric system enables citizens to organize multiple autonomous administrative units at different levels (Ostrom, 2010). Under certain circumstances, polycentric systems promote adaptability and learning capacities, by use of local knowledge and observation of the behavior of similar units through trial-error (Ostrom, 2005).

In Latin America, several authors have described the importance of self-organizing to understand and promote sustainable management practices (Basurto *et al*; 2013; Mendoza-Carranza *et al.* 2013; García Lozano and Heinen, 2016). A recent study of the coast of Peru describes the experience of artisanal fishing communities that have established informal control rules, including limitations on catch volumes and self-monitoring systems, managing to maintain sustainable fisheries over time (Nakandakari *et al.*, 2017). These administration and management rules were not implemented in a single event, but rather through a historical local learning process (Nakandakari *et al.*, 2017). Similarly, as detailed in the following section, Chile has implemented significant transformations in artisanal and industrial fisheries administration regimes (Gelcich *et al.*, 2010).

Promoting and strengthening collective action spaces is a standard diagnosis in self-organizing studies and new governance models for the administration of common pool resources (Basurto *et al*; 2013; Mendoza-Carranza *et al.* 2013; García Lozano and Heinen, 2016). Under certain conditions, collective action is a central element to strengthen the capacity of adapting, learning and self-control, necessary for sustainable management. In this chapter, "governance" is understood as a set of interactions that establish a public/private relationship and state/citizen relationship (Kooiman and Bavinck, 2005; Estévez *et al.* 2019). In this context, governance includes legal principles and norms, as well as judicial negotiations or mediation, including webs of interactions not contained in regulatory frameworks but, for example, linked to citizen movements (Lebel *et el.*, 2016). The establishment of a governance model involves defining how to administrate power among different social actors and their relations with an issue or object of interest (Estévez *et al.*, 2019).

In this chapter we put forward theoretical contributions from the social sciences to help understand collective actions spaces as a condition for sustainable management of common pool resources. The chapter argues that deliberative democracy principles are central to the legitimacy of a governance model (Frame and O'Connor, 2011). The transformation process that began in Chile in 1990 in artisanal fisheries governance is described, exploring the importance of certain principles for long-term sustainability of these systems.

Collective action spaces for sustainable management of common pool resources

Institutional transformations in the administration of common pool resources is analyzed at three levels: the operational space, collective action space and the constitutional space (Ostrom, 2005). These levels are sequential and simultaneous interconnected social spaces, where different outcomes emerge from their interactions. Collective actions spaces are key in the transformation of the rights to use and access common resources. A social space is where individuals with different interests (generally organization representatives) analyze information, execute actions, estimate consequences and address "trade-offs" in order to implement, monitor and enforce compliance of regulations that affect the users of the operational space (McGinnis and Ostrom, 2014; Ostrom, 2005, 2011). In the operational space, individuals exchange goods and services, solve problems, quarrel and extract resources, among other interactions (Ostrom, 2005). For example, in the artisanal fishery, the operational space refers to interactions associated to landings, processing, commercialization and consumption of marine resources. Finally, the third level, defined as the constitutional space, determines regulations and procedures permitted in the collective action space, including who should participate in decision-making and mechanisms to change the rules (McGinnis and Ostrom, 2014). Therefore, the three levels that determine the management of common resources – the operational, collective and constitutional space – join together as a set of rules, coupled to another set of rules (Ostrom, 2005).

In the collective action space, consensus to establish interactions based on reciprocity, confidence and cooperation principles, happens and is renewed based on interactions underpinned by communication action (Habermas, 2004). In other words, participating actors reinterpret their mutual interdependencies and create new agreements and coordinate their actions to reach new agreements. Communication allows a rational individual to provide arguments justifying his actions and opinions to another individual, based on the logical presentation of legitimate individual expectations. However, in the exchange of arguments, individuals consider the common good and not only their expectations of individual utility. This form of interaction contrasts with strategic action where individuals calculate means and ends in search of maximum personal utility.

In the governance of common resources, collective action spaces are fundamental for participatory and sustainable management (Ostrom, 1992). In each collective action space, participants make joint decisions, with the potential for changes in the operational spaces (McGinnis and Ostrom, 2014). In collective action spaces where interaction patterns are repeatedly based on collaboration, confidence, reciprocity conducts and effective practices for the conservation and management of natural resources are stimulated (Agarwal and Ginsberg, 2017; Krasny and Tidball, 2012).

Collective decision-making and deliberative democracy principles

The study and understanding of governance and collective action models has critical implications for the theory of deliberative democracy (Habermas, 1994; Hajer and Wagenaar, 2004). The concept of democracy that arises from the historical modernization process refers to the institutionalization of the use of public reason jointly exercised by autonomous citizens (Habermas, 1994).

Deliberative democracy is a form of government where members in search of a common good make decisions through public deliberation (Cohen, 1997). The concept of deliberative democracy emerges as an ideal model of democracy for institutions that make collective decisions (Cohen, 1997). Theoretically, deliberative policy is based on the capacity to build consensus through the recognition of valid affirmations, subject to criticism through communicative rationality (Habermas, 2004). In other words, to build deliberative dialogue dynamics, individuals in a communication system, through their interactions, recognize other individuals' right to propose and articulate arguments based on their own value judgments, customs and beliefs. On the other hand, individuals recognize that their own arguments may be the subject of criticism, which implies the search for agreements or the need to restructure or strengthen argumentative speech.

Deliberative policy is regulated both by formal democratic spaces (i.e. the parliament), and informal opinion building and political will processes in the public arena (Habermas, 2009). Deliberative policy evolves under a framework of informal relations as or even more relevant than formal spaces (Mansbridge *et al.*, 2012). Deliberative policy is conceived as a network of negotiation processes and agreements, and forms of argument, including moral, ethical and pragmatic discourses, each one based on different assumptions and communication procedures (Habermas, 1994).

The success of deliberative policy depends not only on citizens' collective action but also on the capacity of institutions to formalize procedures and adequate communication conditions (Habermas, 2009). Deliberate democracy procedures for decision-making demand a set of preconditions to obtain reasonable and fair results. Key assumptions include equality in political rights, complete access to information for all participants, legal certainty for agreements and common objectives (Habermas, 2004). In other words, a deliberative process assumes that participant actors have, from a policy perspective, the same rights and duties (i.e. right to vote, opinion, etc.); it also assumes there is an appropriate flow and management of information, not obstructed by individual interests; finally, it assumes there is a space for participants to present their arguments, listen to the other individuals and reach an agreement (Habermas, 2009). Theory and practice emphasize the need for reciprocity mechanisms and conducts to establish trust and legitimacy (Gutmann and Thompson, 1997).

Deliberate democracy has faced theoretical and methodological criticism. One such critique argues there is limited empirical study on real deliberative democracy experiences (Sunstein, 2002). This creates certain level of skepticism regarding the application of deliberative policy in practice. A second critique refers to the results of deliberative policy being unfavorable for vulnerable social groups with fewer abilities for rational argumentative discussion (Sanders, 1997). In other words, the precondition of equality policy for deliberative analysis would not be present in real decision-making processes. Thus, vulnerable social groups have less capacity to undertake logical argumentative operations, and thereby greater difficulty to reach equitable agreements in a deliberative decision-making space. However, this obviates argumentative capacity which organizational leaders often have to articulate group interests in decision-making collective space. In addition, an experienced facilitator who ensures the opportunity to express argumentative discourses of all groups should articulate the deliberative process (Innes and Booher, 2004).

This chapter does not seek to argue that deliberative democracy is the theoretical model that best explains social policy processes. However, it does suggest that deliberative policy principles offer useful conceptual tools to support decision-making processes in collective action spaces in particular.

Institutional transformation for fisheries management: the case of Chile

Study of artisanal fisheries management offers an important model to understand a series of issues surrounding common pool resources (Kittinger *et al.*, 2013) and explore the practical consequences of deliberative democracy principles. In Chile, artisanal fishery includes fishers, divers, shore fishers, seashore collectors and artisanal ship-owners. By 2015, there were close to 95,000 registered artisanal fishers, including more than 20,000 women (SUBPESCA, 2015).

Following the enactment of the General Law on Fisheries and Aquaculture in the 1990s a process of institutional changes was triggered with regards to marine resources management.[1] First, the right to extract and use benthic resources (organisms living on the coastal seafloor) remained exclusively limited to artisanal fishers, excluding industrial fishery fleets and methods – which mostly focus on the extraction of pelagic and demersal fish. Second, the overexploitation of benthic resources, mainly those of high commercial value like *Concholepas concholepas* (loco or abalone) and *Loxechinus albus* (sea urchin), led to the collapse of the artisanal fisheries in the early 1990s (Gelcich *et al.*, 2010). In order to ensure fisheries sustainable management, as of 1994, a special management regime was created for Benthic Resources Management Areas. This regime grants exclusive use rights over a certain coastal and marine territory to artisanal fishers' organizations and exclude any individual or other organization from extracting or using resources in this space. Several studies recognize the innovative and transforming character of the Benthic Resource Management Areas system (Gelcich *et al.*, 2010), reporting positive ecological and biological trends (Castilla *et al.*, 2007). However, the social impacts from these changes in benthic resources management are not clear (Gelcich *et al.*, 2006).

Recently, changes in the General Law on Fisheries and Aquaculture created Management Committees for self-management of marine resources in open areas (Reyes *et al.*, 2017).[2] Unlike Benthic Resources Management Areas which create exclusive use zones for artisanal fisheries organizations, each Management Committee regulates the extraction of a resource (or a group of related resources) in free access areas (benthic crustaceans and pelagic and demersal fish). The Management Committees establish operational regulations for a resource (or group of resources) in a geographical area. Examples would be the Management Committee for the Arauco Gulf for bivalve species such as *Tagelus dombeii* (razor clam) or the Management Committee for Southern Hake for the macro zone (regions X, XI, XIII) which covers demersal fish species such as *Merluccius australis* (southern hake). At present there are 32 Management Committees established; 16 correspond to benthic resources including crustaceans and octopus.

The Management Committees for fisheries are a new public collective action space where actors convene for the purpose of decision-making regarding ecosystem management for fishery resources (Hajer, 2004; Habermas, 2011). This policy constitutes a decentralization strategy for fishery resources management in Chile, granting autonomy to artisanal fisheries organizations in their management decisions depending on the resources and geographical areas. The system promotes a polycentric form of government, with autonomous operational units for the management of marine resources (Gelcich, 2014). This trend towards a polycentric model in the management of marine resources reinforces a process initiated in 1994 with the creation of Benthic Resources Management Areas.

A major task for the Management Committees is to design, implement and monitor a plan for specific resources. The main actions contained in Management Plans relate to the implementation of rules that regulate access to and extraction of resources through control and monitoring strategies, training and research, among other actions. The Management Committees are presided over by an officer from the Sub-Secretariat of Fisheries and Aquaculture formed by a maximum of seven representatives of artisanal fisheries,[3] one representative of processing plants, and representatives of the state agencies as determined by the Management Committee (i.e. Municipalities, Ministry of Environment and National Fisheries and Aquaculture Service, among others).

It is noteworthy that changes to the General Law on Fisheries and Aquaculture in 2013, call for advancing towards a precautionary and ecosystem approach to fisheries and resources management. This has triggered the inclusion of social, economic, biological and environmental objectives in Management Plans. However, the concept of an "ecosystem approach" as included in the law is narrower than that contained in international instruments or regulatory bodies (FAO, 2016).[4] The General Law on Fisheries and Aquaculture does not explicitly incorporate social, economic or environmental considerations – which are an integral part of the ecosystem approach.

Crossroads for deliberative democracy in the management of common pool resources: commitments on agreements and mutual monitoring

A critical challenge which users of common pool resources face is building the capacity to interact based reaching agreements, ensuring cooperation and fulfilling commitments. In other words, how to move from a situation in which users act in isolation, considering individual benefits, to a scenario where coordinated strategies are adopted in order to obtain greater common benefits.

Elinor Ostrom (2005) summarizes the challenges for institutions that manage common pool resources in terms of two great crossroads. First, the problem of commitment and the effort required to structure collective action spaces and establish regulations, would not make sense if users do not commit to the compliance of self-defined rules. Second, in terms of mutual monitoring, if users do not implement self-monitoring practices, the incentives for commitment and compliance are reduced.

Field work and experiments show that when users of a common resource lack communication and learning mechanisms, and/or lack access to collective action spaces in order to change the rules of the operational space, they tend to overexploit and deplete resources (Ostrom, 2011). However, the internal and external variables that promote or hamper cooperative attitudes among artisanal fisheries in Chile, are not clear (Gelcich *et al.*, 2013).

The establishment of Management Committees for marine resources allows research on the practical applicability and theoretical relevance of deliberative democracy principles, to support decision-making in collective action spaces. This requires understanding the institutional transformations produced in the specific public (in a broad sense) domain. This is, the generation of new collective action spaces for artisanal fisheries management. This chapter argues that communicative action principles, deliberative democracy and the ecosystem approach are central components for collective action spaces to transform government models towards an administrative scheme that ensures sustainability in the management of marine resources.

Final reflections

The development of Management Committees is key for the future of artisanal fisheries in Chile. This policy generates changes not only in the rights to administrate these resources, but also in the interactions among social organizations and the state. The policy promotes the creation of collective action spaces as articulation instances for fisheries management. However, the implementation of the policy faces challenges and obstacles, both regulatory and in terms of the administration of institutions and self-management capacity by the artisanal fishers themselves and their representatives. A concerning trend observed in research incudes limited or non-existent budgets to support the policy as well as limited decision-making capacity.

As a general conclusion, it could be said that there is a need to promote and strengthen self-organizing capacities of common pool resource users. Not only is it important to create new collective action spaces which are autonomous in the management of fisheries and resources, but also to develop the capacity of these spaces to establish sound interactions based on deliberative democracy principles such as reciprocity, confidence and cooperation, in pursuit of sustainable management. Strengthening collective action spaces in terms of efficient and transparent decision-making and social legitimacy, can lead to a greater capacity of self-control and self-monitoring in operational spaces. In brief, the mechanism to address the described crossroads in the administration of common pool resources is through collective action based on deliberative democracy principles.

Notes

1 Law 18892, General La won Fisheries and Aquaculture, of September 6, 1991. Available at, www.leychile.cl/Navegar?idNorma=30265.
2 Law 20560 regulates artisanal fisheries and the global catch quota and incorporates benthic management plans, enacted on January 3, 2012. Available at, https://ciperchile. cl/wp-content/uploads/LEY-20560_03-ENE-2012.pdf. Law 20657 modifies the scope sustainability of hydrobiological resources, access to industrial and artisanal fisheries activities and research and inspection. General Law on Fisheries and Aquaculture of February 6, 2013. Available at, www.subpesca.cl/portal/615/w3-article-764.html.

3 For benthic resources – exclusive for artisanal fisheries – the Management Committees exclude participation of industrial fisheries.
4 See principles and concepts in the Convention on Biological Diversity, available at www.cbd.int/doc/publications/ea-text-en.pdf.

References

Agarwal, M., Ginsberg, J.R. (2017), Untangling Outcomes of de Jure and de facto Community-Based Management of Natural Resources. *Conserv Biol.* 31(6): 1232–1246. doi: 10.1111/cobi.12954.

Baur, I., Binder, C.R. (2013), Adapting to Socioeconomic Developments by Changing Rules in the Governance of Common Property Pastures in the Swiss Alps. *Ecology and Society* 18(4):60.

Basurto, X., Gelcich, S., Ostrom, E. (2013), The Social Ecological System Framework as a Knowledge Classificatory System for Benthic Small-Scale Fisheries. *Global Environmental Change: Human and Policy Dimensions* 23:1366–1380.

Bene, C., Macfadyen, G., Allison, E. (2005), *Increasing the Contribution of Small-Scale Fisheries to Poverty Alleviation and Food Security*. FAO Technical Guidelines for Responsible Fisheries. Rome, Italy: FAO.

Bromley, D.C. (1992), *Making the Commons Work: Theory, Practice and Policy*. San Francisco: Institute for Contemporary Studies.

Castilla, J.C., Gelcich, S., Defeo, O. (2007), Successes, Lessons, and Projections from Experience in Marine Benthic Invertebrate Artisanal Fisheries in Chile. In: McClanahan, T. and Castilla, J.C. (Eds.), *Fisheries Management: Progress Toward Sustainability* (pp. 25–42). Oxford: Blackwell.

Cohen J. (1997), Procedure and Substance in Deliberate Democracy. In: Bohman, J., Rehg, W. (Eds.), *Deliberative Democracy: Essays on Reason and Politics* (pp. 407–428). Cambridge: Massachusetts Institute of Technology.

Estévez, R.A., Martínez, P., Sepúlveda, M., Aguilera, G., Rauch, M., Gelcich, S. (2019), Gobernanza y Participación en la Gestión de las Areas Silvestres Protegidas del Estado de Chile. In: Cerda, C., Silva, E. (Eds.), *Dimensión Humana en la Gestión de los Sistemas Naturales* (pp. 381–403). Santiago de Chile, Chile: Editorial Ocho Libros.

FAO. (2016), *Asistencia para la Revisión de la Ley General de Pesca y Acuicultura, en el Marco de los Instrumentos, Acuerdos y Buenas Prácticas Internacionales para la Sustentabilidad y Buena Gobernanza del Sector Pesquero*. Santiago: Organización de Naciones Unidas para la Alimentación y la Agricultura (FAO).

Frame, B., O'Connor, M. (2011), Integrating Valuation and Deliberation: The Purposes of Sustainability Assessment. *Environmental Science & Policy* 14:1–10.

García Lozano, A.J., Heinen, J.T. (2016), Identifying Drivers of Collective Action for the Co-Management of Coastal Marine Fisheries in the Gulf of Nicoya, Costa Rica. *Environmental Management* 57:759–769.

Gelcich, S. (2014), Towards Polycentric Governance of Small-Scale Fisheries: Insights from the New "Management Plans" Policy in Chile. *Aquatic Conservation: Marine and Freshwater Ecosystems* 24:575–581.

Gelcich, S., Hughes, T.P., Olsson, P., Folke, C., Defeo, O., Fernandez, M., Foale, S., *et al.* (2010), Navigating Transformations in Governance of Chilean Marine

Coastal Resources. *Proceedings of the National Academy of Sciences of the United States of America* (PNAS) 107:16794–16799.

Gelcich, S., Edwards-Jones, G., Kaiser, M.J., Castilla, J.C. (2006), Co-Management Policy can Reduce Resilience in Traditionally Managed Marine Ecosystems. *Ecosystems* 9:951–966.

Gelcich, S., Guzman, R., Rodriguez-Sickert, C., Castilla, J.C., Cardenas, J.C. (2013), Exploring External Validity of Common Pool Resource Experiments: Insights from Artisanal Benthic Fisheries in Chile. *Ecology and Society* 18(3):2–19.

Gutmann, A., Thompson, D. (1997), *Democracy and Disagreement.* The Belknap Press of Harvard University Press: Cambridge (originally published in 1996).

Habermas, J. (1994), Three Normative Models of Democracy. *Constellations: An International Journal of Critical & Democratic Theory* 1(1):1.

Habermas, J. (2004), *The Theory of Communicative Action.* Volume 1: Reason and the Rationalization of Society. Polity Press, Cambridge.

Habermas J. (2009), *Between Facts and Norms: Contributions to a Discourse Theory of Law and Democracy.* Cambridge: Polity Press (original German, published 1992).

Habermas, J. (2011), *The Structural Transformation of the Public Sphere.* Cambridge: Polity Press.

Hajer, M. (2004), A Frame in the Fields: Policymaking and the Reinvention of Politics. In: Hajer, M.A., Wagenaar, H. (Eds.), *Deliberative Policy Analysis Understanding Governance in the Network Society* (pp. 88–112). Cambridge: Cambridge University Press, Cambridge.

Hajer, M.A. and Wagenaar, H. (Eds.) (2004), *Deliberative Policy Analysis: Understanding Governance in the Network Society.* Cambridge: Cambridge University Press.

Hardin, G. (1968), The Tragedy of the Commons. *Science* 162:1243–1248.

Innes J.E. and Booher D.E. (2004), Collaborative Policymaking: Governance Through Dialogue. In: Hajer, M.A. and Wagenaar, H. (Eds.), *Deliberative Policy Analysis: Understanding Governance in the Network Society* (pp. 33–59). Cambridge: Cambridge University Press.

Kittinger, J.N., Finkbeiner, E.M., Ban, N.C., Broad, K., Carr, M.H., and Cinner, J.E., *et al.* (2013), Emerging Frontiers in Social–Ecological Systems Research for Sustainability of Small-Scale Fisheries. *Current Opinion in Environmental Sustainability* 5(3–4):352–357.

Kooiman, J., Bavinck, M. (2005). The Governance Perspective. In: Kooiman, J., Bavinck, M., Jentoft, S., Pullin, R. (Eds.), *Fish for Life: Interactive Governance for Fisheries*, pp. 11–24. Amsterdam, Netherlands: Amsterdam University Press.

Krasny, M. and Tidball, K. (2012), Civic Ecology: A Pathway for Earth Stewardship in Cities. *Frontiers in Ecology and the Environment* 10(5):267–273.

Lebel, L., Anderies, J.M., Campbell, B., Folke, C., Hatfield-Dodds, S., Hughes, T.P., Wilson, J. (2006), Governance and the Capacity to Manage Resilience in Regional Social-Ecological Systems. *Ecology and Society* 11 (1):19. Available at, www.ecologyandsociety.org/vol11/iss1/art19/.

Mansbridge, J., Bohman, J., Chambers, S., Christiano, T., Fung, A., Parkinson, J., Thompson, D.F., Warren, M.E. (2012), A Systemic Approach to Deliberative Democracy. In: Parkinson, J., Mansbridge, J. (Eds.), *Deliberative Systems: Deliberative Democracy at Large Scale* (pp. 1–27). Cambridge: Cambridge University Press.

McGinnis, M.D., Ostrom, E. (2014), Social-Ecological System Framework: Initial Changes and Continuing Challenges. *Ecology and Society* 19(2):30.

Mendoza-Carranza, M., Arévalo-Frías, W., Inda-Díaz, E. (2013), Common Pool Resources Dilemmas in Tropical Inland Small-Scale Fisheries. *Ocean & Coastal Management* 82:119–126.

Nakandakari, A., Caillaux, M., Zavala, J., Gelcich, S., Ghersi, F. (2017), The Importance of Understanding Self-Governance Efforts in Coastal Fisheries in Peru: Insights from La Islilla and Ilo. *Bulletin of Marine Science* 93:199–216.

Ostrom, E. (1992), The Rudiments of a Theory of the Origins, Survival, and Performance of Common-Property Institutions. In: Bromley, D.W. (Ed.), *Making the Commons Work: Theory, Practice and Policy* (pp. 293–319). San Francisco: Institute for Contemporary Studies.

Ostrom, E. (2005), *Understanding Institutional Diversity*. Princeton: Princeton University Press.

Ostrom, E. (2010), Polycentric Systems for Coping with Collective Action and Global Environmental Change. *Global Environmental Change: Human and Policy Dimensions* 20:550–557.

Ostrom, E. (2011 [1990]), *Governing the Commons: The Evolution of Institutions for Collective Action*. New York: Cambridge University Press.

Reyes, F., Gelcich, S., Ríos, M. (2017), Problemas Globales, Respuestas Locales: Planes de Manejo como Articuladores de un Sistema de Gobernabilidad Policéntrica de los Recursos Pesqueros. In: Irarrázaval, I., Piña, E., Letelier, M. *Propuestas para Chile: Concurso de Políticas Públicas 2016* (pp. 121–156). Santiago de Chile, Chile: Centro UC Políticas Públicas, Pontificia Universidad Católica de Chile.

Sanders, L.M. (1997), Against Deliberation. *Political Theory* 25:347–376.

Schlager, E., Ostrom, E. (1992), Property-Rights Regimes and Natural Resources: A Conceptual Analysis. *Land Economics* 68:249–262.

Sunstein, C.R. (2002), The Law of Group Polarization. *Journal of Political Philosophy* 10:175–195.

SUBPESCA. (2015), *Anuario Estadístico de Pesca 2015: Subsector Pesquero Artesanal*. Servicio Nacional de Pesca y Acuicultura (SUBPESCA). Chile.

12 The legal protection of surf breaks

An option for conservation and development

Bruno Monteferri, Christel Scheske and Manuel Ruiz Muller

Introduction

Surf breaks are areas which provide suitable waves for surfing and other related sports. The shape of the coast, sea-bed conditions and direction and strength of swells and wind allow for surfing, bodyboarding, windsurfing, paddle-boarding, body-surfing and kite-surfing to take place.

Surf breaks are part of natural infrastructure, a concept that seeks to highlight the value and functions of natural spaces and ecosystems to provide benefits for human populations. Unlike other sports such as football, that requires the construction of fields and stadiums, for surfing one only needs to care for the surf breaks and guarantee adequate access and basic and simple infrastructure for public use.

Surf breaks provide opportunities for tourism, recreation, aesthetic inspiration and the construction of national or regional identity and can be seen as true cultural ecosystem services. Cultural ecosystem services are nonmaterial benefits people obtain from ecosystems through spiritual enrichment, cognitive development, reflection, recreation and aesthetic experiences (Millennium Ecosystem Assessment 2003).

The benefits aquatic sports provide for physical and mental health have been well documented (Nichols, 2014). Furthermore, in a context where the impacts and degradation of marine ecosystems are increasing and the human population is suffering from the so-called "nature deficit disorder" (Louv, 2008), governments and civil society need to consider surfing and other aquatic sports because of their potential in helping citizens to give more value and commit to caring for marine environments. Scheske *et al.*, 2019 (*in press*) refers to several social psychology studies showing that marine sports such as surfing can increase interest and involvement in marine conservation. The authors also highlight the work of more than ten civil society organizations created by surfing groups to protect surf breaks and coastal and marine ecosystems, starting at the local level (i.e. Fundación Punta de Lobos in Chile) and moving to thousand-member organizations at the global level (i.e. Surfrider Foundation).

Surf breaks have a high economic value. This is clearly demonstrated in new research on "surfonomics" (Lazarow *et al.*, 2009; Thomas, 2014). For example, the Gold Coast in Australia generates around US$180 million a year from visiting surfers (Lazarow, 2008). It is estimated that the value of a house in Santa Cruz, California, near a surf break, is US$106,000 more than a similar house far from the break (Scorse *et al.*, 2015).

Preliminary studies also show that there are many cases where surf breaks overlap with areas of high biodiversity, hence the convergence of interests of marine conservation and surf break protection. For example, in 2018, the NGOs Conservation International and Save the Waves Coalition formed a new alliance and published a map that identifies areas where biodiversity "hotspots" and iconic breaks around the world overlap (Conservation International, 2018). Scheske *et al.* (*in press*) mentions the examples of Peru and Chile where emblematic breaks are a part of coastal and marine landscapes with high biodiversity value such as the Illescas Peninsula in Peru or Punta de Lobos in Chile. Marine conservation can contribute to protecting surf breaks and surf break protection while also benefitting conservation.

Threats to surf breaks

Many surf breaks around the world are threatened both by natural events and direct and indirect human causes (Save the Waves, n.d.). Coastal development projects are particularly concerning as they mostly overlook or understate the potential impact of infrastructure such as seaports, breakwaters, piers, underwater pipelines and unplanned developments on coastal areas (Corne, 2009). These projects can alter the sedimentation cycles on which surf breaks depend to maintain their quality or wave formation and trajectory. Natural events can also have dramatic effects. For instance, the eruption of the Kīlauea Volcano in Hawai'i in 2018, was responsible for the disappearance of several surf breaks (Diskin, 2018). Likewise, tsunamis and earthquakes in Indonesia have modified the sea floor (Brokensha, 2012) and several surf breaks almost disappeared or were gravely affected. Finally, sea rise from climate change is also generating various effects which are altering the quality of surfable waves (Harley *et al.*, 2006; Reineman *et al.*, 2017).

It is not only surf-related sports that are being affected by these threats but also the ecosystem benefits they provide and, in certain cases, the biodiversity of the area. The threats to surf breaks have prompted answers from civil society, particularly the surfing community, and also from governments and international organizations (Scheske *et al.*, in press). In the following section we explore some legal tools and strategies that are being used in various parts of the world to legally protect the breaks.

Surf breaks as legally protected subject matter

For an object or subject to be protectable under the law, it first has to be legally recognized as such. For example, when water, rivers and forests are recognized as natural resources or part of the natural heritage, legal norms are applied to regulate their access, use, management and conservation, including through administrative and criminal provisions to ensure their protection. Second, the protection granted must respond to concrete objectives which define the parameters and scope of protection. Surf breaks have historically been "invisible" for legal frameworks and have only recently started to be recognized as concrete legal subject matter. Sometimes, surf breaks were protected indirectly through social recognition or through legal or administrative measures pertaining to constitutional rights such as the right to a healthy environment or the right of access to a public area. In other cases, their protection has been possible through measures intended to safeguard and protect coastal borders or marine ecosystems. Insofar as the breaks are part of coastal and marine ecosystems, a first approach to their protection was through strategies for the conservation of such ecosystems. However, their specificity was hardly, if at all, part of a conscious effort to protect them.

For example, landscape in Peru is considered a natural resource when it is in some way exploitable for economic purposes. Therefore, for countries which similarly treat land and landscapes in general, one option is to consider surf breaks as natural resources and, hence a part of the coastal and marine landscape susceptible to being exploited and used for economic purposes. In countries with legal frameworks for ecosystem services, surf breaks could be protected by justifying the cultural and other services they provide. Furthermore, in some countries surf breaks could be recognized as a tourism resource that should be valued and protected. These are just a few indirect ways in which surf breaks may be protected.

Ideally, however, protecting surf breaks requires their recognition as legal subject matter and creating a specific legal framework that responds to their distinct characteristics. This could be done through legal recognition of surf breaks and providing a legal/technical definition. A particular law or regulation may include specific references regarding scope of protection (i.e. the break per se and/or surrounding areas), institutional responsibilities (i.e. what institutions are in charge of the management, monitoring, maintenance, etc.) and financing (i.e. how surf breaks will be maintained and interventions to guarantee their preservation over time funded), among others. Regardless of whether it is a single law or a set of laws and regulations which add to surf break protection, it is important to guarantee an adequate, effective and efficient level of protection through sound and reasonable measures.

The protection of breaks through ad hoc laws

At present, Peru is the only country with a national law in place specifically developed to protect surf breaks. As a result of significant threats to emblematic surf breaks such as La Herradura (Lima) and Cabo Blanco (Piura) during the 1990s, the Law for the Preservation of Surf Breaks was approved in 2001.[1] This law defines surf breaks as Natural Patrimony and they are recognized as state property. It mandates the Peruvian Navy to create a national register for their protection: the National Register for the Protection of Surf Breaks. A regulation added to the law in 2013 specified the administrative procedures and requirements to register surf breaks with the National Register.[2] To protect a surf break, the National Surfing Federation is required to submit an application to the Peruvian Navy – which has the competence of granting rights to aquatic areas – which must include the name of the surf break, location, geographic position and universal coordinates, a map of the area(s) to be protected, descriptive technical profile and bathymetry studies that justify the existence of a surf break and that it is appropriate for surfing and related sports.

If the application is approved, the Peruvian Navy issues a Directorate Resolution approving protection and inclusion of the surf break in the National Register. By including a surf break in the National Register, the Peruvian Navy can no longer grant other use rights to the same aquatic area. This generates legal restrictions on other forms of use in the area, mainly related to infrastructure, oil and gas exploration, aquaculture concessions, protecting sedimentation processes and the integrity of the marine landscape. The law also requires that activities or projects in the "adjacent zones" of a surf break – a maximum of one kilometer lengthwise of the coast, measured from both sides of the break – must include damage prevention or mitigation measures.

Registration of a surf break costs between US$5,000 to US$8,000 given the technical profile and bathymetry studies of the area that need to be presented to the Peruvian Navy. In practice, registrations in the National Registry started when the Peruvian Society for Environmental Law, in alliance with the National Surfing Federation launched the "Do it for Your Wave" (HAZla por tu Ola) campaign. As a result, this campaign has triggered a real and effective social movement to protect waves in Peru. To date, 33 surf breaks have been formally registered and are protected thanks to contributions from thousands of citizens, companies, foundations and municipalities who donated money to cover the costs of preparing the technical profiles for each surf break. Furthermore, surf break protection has been made possible through collaboration between government and civil society. This truly participatory process has given legitimacy and social support to surf break protection.

The effectiveness of the law and the National Register has already been put to the test in cases of poorly planned coastal infrastructure. For example, the wave in Cabo Blanco was protected from the construction of a new fishing dock that in the original plans would have significantly affected the waves' quality and function. As a result of a multilateral dialogue process among fishers, surfers and public entities, actors (including local fishers) acknowledged that changes in dock construction plans were required to reduce potential impacts to the Cabo Blanco wave. Another case is Huanchaco, a long-time surfing destination for tourists and known worldwide for its *"caballitos de totora"*, which have been traditionally used by artisanal fishers to cross and surf waves for thousands of years (Meza *et al.*, 2015). This specific surf break in Huanchaco – the second to be formally registered – was threatened by a proposal to construct nine breakwaters along the beach as a measure to avoid the process of coastal erosion caused by the breakwater constructed kilometers away in the Port of Salaverry. The initial development proposal was modified and reduced to three docks, thanks to members of the Huanchaco World Surfing Reserve who used the law as one of their legal arguments to protect the Huanchaco surf break.

It is worth highlighting that in a country like Peru, where there are neither marine spatial planning processes nor integrated management of coastal zones, and decisions often made with poor inter-sectorial coordination, a site protected by the law may help to reduce the threats but is not enough to avoid other effects on the surf break. It is crucial to have groups of surfers and civil society actively organized to defend surf breaks and monitor compliance with the law and its regulation. The advantage of having the law in place is that it offers civil society and advocacy groups a solid legal tool to with which to address threats like environmentally risky development projects.

The protection of breaks in the framework of coastal zone management policies

Coastal and marine planning and conservation strategies in general can also play a key role in the protection of surf breaks. There are several examples across the world of how countries and local jurisdictions develop plans and strategies to protect, in general, their coastal and marine zones and include specific references to surf breaks therein.

In New Zealand, the government actively protects surf breaks by explicitly including them in costal-marine management plans. Under the framework of the Resource Management Act of 1991, the New Zealand Coastal Policy Statement specifically refers to the protection of surf breaks of national importance.[3] The act mandates that surf breaks must not be affected by coastal development activities, while the Policy Statement determines that the access, use and enjoyment of surf breaks must not be negatively affected by activities related to coastal development. The Policy Statement has served to protect 17 breaks. These legal instruments help prevent coastal and marine development (e.g. docks, marinas and piers, housing and infrastructure in general) from having adverse impacts on surf breaks. Similar to the Peruvian case, in New Zealand there are organizations such as the Surfbreak Protection Society that have actively participated in the implementation of the act and Policy Statement. They managed to prevent several dangerous projects in coastal areas, and ensured that surf break protection was included in planning and policy-making agendas at different levels.

In Australia, the Gold Coast City Council in the state of Queensland enacted the Gold Coast Surf Management Plan in 2015. It is the first planning instrument of its type, where a local government and the community develop an integrated plan detailed for surfing and surf breaks. This is an example of how actions and measures to guarantee surf break protection can be incorporated into strategic planning processes and instruments.

Finally, in the United States of America, the California Coastal Commission recognizes sensitive coastal areas as areas that have significant recreational value – surf breaks can qualify under this general category.[4] The California Coastal Commission has powers that help in planning the use of coastal and marine zones under county jurisdictions which are responsible for undertaking actions to: prevent sea pollution, guarantee public access to waves and avoid constructions that affect the landscape and ecosystem services. Likewise, the Hawaii Coastal Zones Management Program includes references that indirectly contribute to surf break protection. In this case, specifically facilitating the right of access to beaches and enjoyment of spaces, as a way of encouraging a culture of respect and appreciation for the sea, beaches and waves.

Surf break protection through the creation of protected areas

In their analysis regarding possibilities of integrating marine biodiversity conservation with surf break protection, Scheske *et al.* (in press) analyze the relationship between protected areas and surf breaks. Protected areas are clearly defined geographical spaces, recognized, dedicated and managed to achieve the long-term conservation of nature with associated ecosystem services and cultural values (Dudley, 2008). Since the first protected area was established at the end of the nineteenth century in the United States of America, protected areas have consolidated as the main global biodiversity conservation tool. In spite of potential shared goals between marine conservation and surf break protection, there are a few existing cases where protected areas have been specifically created to preserve surf breaks.

Scheske *et al.* (*in press*) argue that some surf breaks and surrounding areas may meet criteria for consideration and recognition as protected areas. According to their analysis, there are four categories particularly appropriate for surf break protection under IUCN categories for protected areas (see Table 12.1). Category III for cases where surf breaks are the main purpose of protection and Categories II, V and VI when surf breaks are part of a wide range of coastal and marine features that are considered of importance. Since surf break protection involves access to and use of sites by surfer, beach goers and sportspeople in general, they qualify under IUCN categories and zoning areas that allow recreational activities. Surf breaks' strong visitor appeal is an excellent opportunity for the implementation of sustainable nature-based tourism approaches while generating funding for these protected areas.

Until recently, surf breaks have generally not been considered when creating protected areas nor been part of planning processes as a specific value to be reckoned with (Sheske *et al.*, in press). One of the few cases where a goal of the protected areas has been to protect surf breaks is the Marine Sanctuary of the Municipality of Natividad in Chile, which was explicitly created to

Category II – National Park: The "National Park" category is appropriate for surf break protection that is part of a larger size protected area reserved to maintain ecological integrity at an ecosystem level. The goals of National Parks include promoting education, recreation and contributing to local economies through sustainable tourism. Surfing can be a recreational and educational experience promoted at National Parks, implementing regulations and zoning measures to minimize and mitigate the impacts for ecosystems protected by visiting surfers. For example, G-Land in East Java, Indonesia, is part of the Alas Purwo National Park. G-Land is one of the most iconic waves in Indonesia. Only the construction and operation of three surfing eco-lodges has been allowed, with a limit of surfers authorized to enter at the same time.

Category III – Natural Monument or Feature: The category "Natural Monument" is appropriate for the protection of emblematic surf breaks when they are not part of a broad landscape/seascape that needs protection. The break in itself and immediate surrounding characteristics such as rock formations could be the object for conservation under a protected area. In this case, the value of the surf break is based on values attributed by surfers, as well as related actors that depend on surfers (i.e. tourism operators, restaurants, surfing schools). The Lobitos break in northern Peru is considered one of the most perfect waves the country offers. It was included on the list of potential land and sea pilot sites to create a natural monument system (SPDA, 2016). The designation of a surf break and surrounding beaches as a natural monument would not only contribute to increasing the area's visibility and "presence", but also help to guarantee implementation of management measures and regulations concerning limited infrastructure development, limits to the use of motorized vehicles on beaches and dune conservation, all of which play a key role for beach sanding processes on which the waves depend.

Category V – Protected Landscape/Seascape: Category V is appropriate for the protection of surf breaks that are part of a broader terrestrial coastal and marine landscape that is worth protecting, mainly for its landscape value and sustainable interaction among human beings and nature of the zone. In this case, the break need not be the main focus to create a protected area. However, it must be explicitly mentioned to be included under management measures. For example, this would allow the control of surf tourism, offering an income for the protected area's management and control. The Monterrey Bay National Marine Sanctuary in California is considered a Category V protected area, covering 1.5 million hectares, where a number of uses are possible. There are two prominent surf breaks within the sanctuary: Mavericks, an iconic big surf break and Steamer Lane in Santa Cruz. Surfing is explicitly addressed in the sanctuary's management plan, for example, by regulating the use of personal motorized watercraft (generally jet skis) particularly used for big wave surfing, in order to minimize the impacts on wildlife (National Marine Sanctuaries, 2008).

Category VI – Protected area with sustainable use of natural resources: This category is focused on large areas that combine conservation with the sustainable use of natural resources. As in the case of Category V, surf breaks are part of a set of broader conservation objectives. Surfing and tourism are different types of sustainable use in the area, such as fisheries regimes managed at the local level. A successful example is the Paracas National Reserve in Peru of 217,594 hectares, a Category IV marine protected area established in 1975. The reserve includes a valuable surf break on San Gallan Island, accessible only by boat. Throughout its history, the break was part of a strict protected zone, where surfing was officially banned (INRENA, 2002). However, surfers continued to visit the zone and even organized yearly surfing competitions (W. Wust, 2018, pers. comm., November 2018). Around 2015, dialogues were initiated between surfers and the management committee of the reserve, to legalize and regulate surfing in San Gallan, to value surfing for its potential income for the reserve, and at the same time ensure the impact on wildlife is kept at a minimum. The process was a success: in 2016 the new management plan for the reserve changed the norms on zoning around the break to allow its sustainable use (SERNANP, 2016). Another significant step is that now the break is explicitly mentioned as an asset for the reserve, its access is regulated by authorized tourism operators (SERNANP, 2016).

promote surfing and kite surfing and other recreational purposes (Ilustre Municipalidad de Navidad, 2008).

However, there are no comprehensive studies which assess the presence of surf breaks within the limits of protected areas nor how are they regulated. Regulation is important to ensure that tourism and sports such as surfing, scuba diving and recreational use of motorboats have as little negative impact on coastal and marine environments as possible, through appropriate management (Davenport and Davenport, 2006). In areas of high biodiversity, it is critical that regulations are implemented to reduce the impacts generated from surfing. These impacts are produced by people walking on reefs, cars driving through bird and turtle nesting areas and jet ski watercraft driving through important areas for marine fauna, generating oil residue and disturbing marine life and environments.

Other measures that contribute to surf break protection

In addition to specific legal frameworks for surf break protection, protected areas and the development of strategic planning instruments that make reference to coastal and marine zones and in some cases surf breaks, there are other alternatives that contribute to surf break protection.

World Surfing Reserves. World Surfing Reserves (WSR) is a concept created by the Save the Waves Coalition as a first step for local communities to organize, generating a sense of identity and pride for their waves, and adopt strategies to protect them and value them – through the legal category or tool that best applies to the specific situation. There is a process to apply and be recognized as a WSR, and there are already 10 recognized WSRs in countries like Australia, Chile, Mexico, Peru and Portugal, among others (Save the Waves, n.d.). World Surfing Reserves have no strict legally binding effect, but help to increase public awareness and recognition of the contributions surf breaks make to the economy and local development, especially to strengthen ties and organizational capacities of local communities. Furthermore, the alliance with a global platform such as Save the Waves Coalition provides local groups with more visibility in the media, capacity for influence and access to international support in the face of threats. The objective of Save the Waves Coalition is for each WSR to evolve and convert, when the right time comes, into a duly protected site with legal and administrative safeguards. This requires working within the legal and regulatory frameworks of each country. For example, in the case of Huanchaco (Peru), the break is recognized as a WSR and at the same time is formally included in the National Register for the Protection of Surf Breaks. The local stewardship council of the Huanchaco WSR also triggered the process to legally protect the surf break through the Law for the Protection of Surf Breaks when the waves in Huanchaco were threatened.

International natural patrimony. Internationally, the United Nations Education, Scientific and Cultural Organization (UNESCO) administers the Convention concerning the Protection of World Cultural and Natural Heritage (1972), that recognizes sites of cultural and natural importance. There are several examples of Natural Heritage Sites that include world-class surf breaks under their geographical extension, including: the Ujung Kulon National Park in Indonesia where the legendary One Palm Point break is located; the Guanacaste Conservation Area in Costa Rica where emblematic waves are found in Junquillal, Playa Negra, or Roca Bruja; and the Aldabra Attol in Seychelles where you can find dozens of world-class breaks. Although the declaration of an area as a World Heritage Site does not provide legal protection for the space, it does put pressure on countries in charge of their control, to guarantee their adequate management and care, otherwise they may lose international recognition.

Biosphere reserves. In addition to Natural Heritage Sites, UNESCO recognizes Biosphere Reserves to help protect terrestrial and marine areas that contain flora and fauna of special interest for science, and where solutions to reconcile biodiversity conservation with sustainable development are promoted. There is a World Network of Biosphere Reserves, where research, exchange of knowledge and experiences are promoted to improve relations between people and nature. The UNESCO Man and Biosphere Programme (MAB) coordinates these efforts globally and each country creates instances for their management. Under Biosphere Reserves, actions related to conservation and the development of surf breaks could be promoted. A list of breaks that are part of a Biosphere Reserve has yet to elaborated.

National cultural sites. In some countries like Australia, the United States of America and South Africa, laws have been enacted recognizing cultural and emblematic dimensions of certain coastal and marine spaces. Often this is the result of the presence of emblematic surf breaks. For example, in San Diego, Santa Cruz and Los Angeles in California, the National Historic Preservation Act has been used to protect – through a National Register – historic sites such as Malibu Beach Town or Windansea Surf Shack, inseparably linked to the role surf breaks have played in these areas to forge cultural and historic links. In Hawai'i, the small town of Hale'iwa, north of O'ahu, was recognized in 1984 as a State Historic, Cultural and Scenic District, partly because of its emblematic surf breaks. In 1992, the High Court of Australia decided to recognize customary rights to the land of indigenous peoples in the Murray Islands in the Torres Strait, indirectly generating the protection of diverse breaks also surfed by aboriginal peoples and thus consolidating local culture.

National security. Some emblematic breaks are protected under national security norms or are in military installations as in the case of the Diego Garcia atoll under the jurisdiction of the U.S. Marines or Playa Ñave in the District of Chilca, Peru, under the patrol of the Peruvian Air Force. There are dozens of surf breaks under military control in Morocco and Hawaii (Coleman, 2014). Although public access to these sites is extremely difficult, with some regularity either legally or violating security, surfers enjoy these breaks at their own risk, as part of the adventure associated with marine sports such as surfing, windsurfing and others.

TURFs. There are also innovative and creative ways to protect surf breaks. In Chile, although they do not have a specific legal framework for surf break protection, the NGO Fundación Rompientes studies possibilities in the context of Territorial Use Rights in Fisheries (TURF) Scheske *et al.* (in press). Through the Management and Exploitation Areas of Benthic Rersources, the Fisheries Sub-Secretary of Chile assigns exclusive rights to exploit benthic resources from areas on the seabed, to fisheries organizations that are responsible for the implementation of resource management plans and creation of monitoring measures to fight against illegal extraction. The Fundación Rompientes is exploring the possibility of including agreements with the fishermen on surf break protection within these areas, providing new economic opportunities for local communities through low-impact surfing tourism on the basis of use restrictions in their management plans.

The protection of surf breaks through OECMs under the framework of the Convention on Biological Diversity

In the world of conservation, the role of Other Effective Area Based Conservation Measures (OECMs) is currently being discussed. These include conservation methods in places that are not protected areas per se but geographical spaces which are managed and have effective results for biodiversity conservation (IUCN-WCPA, 2018).

Scheske *et al.* (in press) argue that some of the measures described previously for surf break protection comply with the criteria for OECMs. The recognition of OECMs allows countries to increase the amount of protected conservation spaces reported to the Convention on Biological Diversity and comply with the Aichi Targets, mainly Target 11 under which countries committed to protecting at least 17 percent of terrestrial and inland water areas and 10 percent of coastal and marine areas worldwide. In the past, countries could only recognize protected areas – but since the approval of Decision 14/8 of the Conference of the Parties to the CBD in November 2018, OECMs can now be included in the accounting of protected areas by each country in order to meet the target. Thus, in cases where a legal mechanism for surf break protection is also effective for the protection of biodiversity of the site, an important incentive is created for the government to promote and strengthen this mechanism, as it will also help with compliance with international commitments (Scheske et al., in press).

Conclusions and final reflections

This chapter has sought to explore existing alternatives for surf break protection. Strengthening the ties between surfing and marine conservation represents an opportunity for conservation groups to secure support for marine conservation initiatives. For example, as surfers are exposed to marine pollution, they tend to get involved in beach clean-up initiatives and campaigns to reduce marine contamination, this being a priority for organizations such as Surfers Against Sewage (SAS) of England, Surfrider Foundation internationally, HAZla por tu Ola in Peru, among others. Additionally, citizen support can be increased for the creation of protected areas that include surf breaks, while implementing good practices to prevent and mitigate possible impacts by surfers on the ecosystems and species, such as limiting the use of cars on beaches where birds are nesting or walking on coral reefs. In this regard, a recent report from IUCN emphasizes the opportunities from associating sports with conservation and provides some tools to minimize and mitigate the negative impacts the sport has on biodiversity (IUCN, 2018).

Where possible and appropriate, governments and civil society must consider surf breaks in protected areas planning, particularly Categories II, III, V and VI. As Scheske *et al.* (in press) conclude, more studies are needed to identify sites where surf breaks converge with high coastal-marine biodiversity areas, for which surfers need the conservation community. A further study needed is an international comparative review of different existing protection mechanisms for surf breaks, and how they could be integrated with marine conservation targets. Clearly, creating alliances between conservation organizations and surf groups is key to benefit from the opportunities – and address the challenges – that exist in the crossroads between surf break protection and marine conservation.

Meanwhile, WSR mechanisms, specific laws like the Law for the Preservation of Surf Breaks in Peru and other management policies for the coastal border, are viable and real options to protect surf breaks and guarantee that they continue to provide multiple cultural, economic and socio-environmental benefits.

Although practices and realities among countries vary considerably, an analysis of the tools mentioned in this chapter highlights some basic elements to support surf break protection mechanisms.

1   Subject matter: surf breaks must be recognized as legal subject matter in a law or regulation to ensure appropriate levels of protection. This should include the surf break per se, considering factors on which the waves depend to maintain their quality, such as sedimentation, bathymetry and wave action. It is important to consider adjacent zones where special attention is needed to avoid interventions that might affect the surf break.

2   Limitations to activities: a legal framework must specify the types of activities that are limited within the designated area such as a protected surf break, and what additional steps must be taken to prevent and mitigate impacts. The forms of protection provided by the legal framework must be clear. For example, infrastructure development projects in adjacent zones to the protected surf break should present environmental impact studies to demonstrate that the surf breaks' functions and services will not be affected.

3   Responsibilities and governance: the legal framework must clearly delimit responsibilities of competent authorities for the compliance with laws and regulations for surf break protection. Likewise, coordination mechanisms are needed.

4   Consequences of the damage to surf breaks: consequences of non-compliance must be dissuasive enough (e.g. through sanctions). Legal frameworks need to include sanctions and specify crimes, which are adequately controlled and supervised by authorities.

5   Active citizens: in most cases where surf breaks in different countries are effectively protected, it becomes evident that citizen pressure is important for the legal protection of breaks, and once protected by law, to ensure that the norms are complied with. These actors can be NGOs, companies (such as tourist hotels for surfers), groups of surfers and specific individuals. In the Peruvian case, a commission officially recognized by the National Surfing Federation was formed and is responsible for enforcing surf break protection. More than a formal recognition, it is important that legitimacy in the processes is established and communication of threats is effectively transmitted to a wide range of relevant actors.

In a context of major pressure on coastal and marine ecosystems, surf break protection is a goal that must be considered transversal to sustainable development of coastal areas. To this effect there needs to be balanced interests from large groups of actors, particularly coastal communities and sportspeople, industries and other stakeholders. Mechanisms that provide legal certainty to support initiatives for the protection of these spaces must be a priority for governments and civil society.

Notes

1   *Ley de Preservación de las Rompientes Apropiadas para la Práctica Deportiva*, Ley No. 27280 (2000). Congreso de la República. Peru.
2   *Reglamento de la Ley de Preservación de las Rompientes apropiadas para la Práctica Deportiva*, Ley No. 27280 (2013). Decreto Supremo No. 015–2013-DE. Peru.
3   New Zealand Coastal Policy Statement, 2010, Policy 16, Surf Breaks of National Significance.
4   Article 3, Recreation, Section 30220 Protection of Certain Water-oriented Activities. California Coastal Act. See, www.coastal.ca.gov/coastact.pdf.

References

Brokensha, C. (2012). The Mentawai Islands: The Rise and Fall of a Surfing Wonderland. Available at, www.swellnet.com/news/surfpolitik/2012/11/15/mentawai-islands-rise-and-fall-surfing-wonderland.

Coleman, J. (2014). Exclusivity: The Price of Uncrowded Perfection. *The Inertia*. Available at, www.theinertia.com/business-media/exclusivity-the-price-of-uncrowded-perfection/.

Conservation International (2018). Conservation International and Save the Waves to expand the World Surfing Reserves. Available at, www.conservation.org/NewsRoom/pressreleases/Pages/Conservation-International-and-Save-the-Waves-to-Expand-World-Surfing-Reserves.aspx.

Corne, N. P. (2009), The Implications of Coastal Protection and Development on Surfing. *Journal of Coastal Research, 25*(2) 427–434.

Davenport, J., and Davenport, J. L. (2006), The Impact of Tourism and Personal Leisure. Estuarine, Coastal and Shelf Science, 67(1–2), 280–292. Available at, www.sciencedirect.com/science/article/pii/S0272771405003999.

Diskin, E. (2018), The Kilauea Lava is Destroying Popular Hawaii Surf Spots. Available at, https://matadornetwork.com/read/hawaii-surf-spots-kilauea-eruption/.

Dudley, N. (Ed.). (2008), *Guidelines for Applying Protected Area Management Categories.* Gland, Switzerland: IUCN.

Harley, C. D., Randall A. H., Hultgren, K. M., Miner, B. G., Sorte, C. J., Thornber, C. S. and Williams, S. L. (2006), The Impacts of Climate Change in Coastal Marine Systems. *Ecology Letters, 9*(2), 228–241.

Ilustre Municipalidad de Navidad. (2008). Santuario de la Naturaleza Marino Las Brisas de Navidad. Navidad, Chile: Ilustre Municipalidad de Navidad.

INRENA. (2002), *Reserva Nacional de Paracas: Plan Maestro 2003–2007.* Lima, Peru: Instituto Nacional de Recursos Naturales (INRENA).

IUCN. (2018), *Sport and Biodiversity.* Gland, Switzerland: IUCN.

IUCN-WCPA. (2018), (Draft) *Guidelines for Recognizing and Reporting Other Effective Area-Based Conservation Measures* (1st ed., Vol. 10). Gland, Switzerland: IUCN.

Lazarow, N. (2008), *A Socio-Economic Study of Recreational Surfing on the Gold Coast.* Griffith Centre for Coastal Management Research Report No 89. Gold Coast, Griffith University.

Lazarow, N. (2009), Using Observed Market Expenditure to Estimate the Value of Recreational Surfing to the Gold Coast, Australia. *Journal of Coastal Research,* SI 56 (Proceedings of the 10th International Coastal Symposium), 1130–1134. Lisbon, Portugal.

Louv, R. (2008), *Last Child in the Woods: Saving our Children from Nature-Deficit Disorder.* Chapel Hill: Algonquin.

Meza, R., Tramontana, O., and Pardo, C. (2015), *5,000 Años Surcando Olas. La Historia de la Tabla en el Perú.* Lima, Perú: Wust Ediciones.

Millennium Ecosystem Assessment (2003), www.millenniumassessment.org/es/.

National Marine Sanctuaries. (2008), *Monterey Bay National Marine Sanctuary: Final Management Plan.* Monterey, USA: NOAA Ocean Service.

Nichols, W.J. (2014), *Blue Mind. The Surprising Science that Shows Us How Being In, On or Under Water can Make you Happier, Healthier and More Connected and Better at What you Do.* New York: Little, Brown.

Reineman, D. R., Thomas, L. N., and Caldwell, M. R. (2017), Using Local Knowledge to Project Sea Level Rise Impacts on Wave Resources in California. *Ocean & Coastal Management, 138*, 181–191.

Save the Waves (n.d.). World Surfing Reserves. Retrieved from: www.savethewaves.org/programs/world-surfing-reserves/.

Scarfe, B. E., Healy, T. R., Rennie, H. G. and Shaw, T. (2009). Mead Sustainable Management of Surfing Breaks: Case Studies and Recommendations. Journal of Coastal Research, 25(3): 684–703.

Scheske, C., Arroyo Rodriguez, M., Buttazoni, J.E., Strong-Cvetich, N., Gelcich, S., Monteferri, M., Rodríguez, L.F., and Ruiz, M. (2019, in press). Surfing and Marine Conservation: Exploring Surf-Break Protection as IUCN Protected Area Categories and other Effective Area-Based Conservation Measures. *Aquatic Conservation: Marine and Freshwater Ecosystems.*

Scorse, J., Reynolds, F., and Sackett, A. (2015). Impact of Surf Breaks on Home Prices in Santa Cruz, CA. *Tourism Economics, 21*(2), 409–418. https://doi.org/10.5367/te.2013.0367.

SERNANP. (2016), *Reserva Nacional de Paracas: Plan Maestro 2016–2020.* Lima, Perú.

SPDA. (2016), *Estudio Legal de Factibilidad para el Establecimiento de una Red de Monumentos Naturales en el Perú.* Lima, Perú: Sociedad Peruana de Derecho Ambiental (SPDA).

Thomas, G. (2014), *Surfonomics Calculates the Worth of Waves. Washington Post,* August. Available at www.washingtonpost.com/surfonomics-quantifies-the-worth-of-waves/2012/08/23/86e335ca-ea2c-11e1-a80b-9f898562d010_story.html?utm_term=.d13b30f01cf6.

Epilogue

Manuel Ruiz Muller, Rodrigo Oyanedel and Bruno Monteferri

This compilation of chapters has enabled us to visualize the viewpoints of well-known authors on a wide range of issues that integrate the discussion on seas, coasts and fisheries. Given the diverse and wide range of perspectives covered by the chapters it is difficult to establish similarities and equivalencies among the situations and contexts described. But there are concerns and issues raised which crosscut the different contributions.

A common element highlighted by the chapters is the undeniable pressure and growing deterioration of marine and coastal ecosystems and fisheries, mainly as a result of human-induced action. Unplanned infrastructure, over-fishing both at the artisanal and industrial level, marine pollution, informality and climate change – yet to be quantified with precision – gradually but significantly impact ecosystems and their resources and, therefore, coastal communities and urban center livelihoods.

As a result of this situation, several of the chapters emphasize institutional issues as a key enabling factor to create and streamline policies for fisheries planning, overfishing, protection of migratory species and marine protected areas, among others. Although with differences between Chile, Mexico and Peru, it is clear that there is a need to consolidate and further strengthen institutions responsible for policy-making and managing, operating, monitoring and inspecting marine and coastal ecosystems. This is where the recipe to construct healthy and resilient marine ecosystem resides. The effects of strong institutions and processes positively impact awareness and enhance the actions of stakeholders and civil society.

Planning and an ecosystem approach must be added as a form of incorporating diverse variables to analysis of options and management responses. Implementation of integral measures helps avoid solutions based on remedial actions founded on a single element, a target species, a certain fishery or a single activity linked to the space under analysis and planning. Although the ecosystem approach entails distinct challenges at the implementation stage, the need to integrate it should be a sufficient incentive to overcome any real or perceived barriers and challenges. The integrity of the marine and coastal space demands bold, comprehensive, articulated and consensual solutions among different actors and stakeholders that converge in these spaces and on their resources. Diverse methodological approaches such as a focus on common goods and contributions from the economy and social sciences, helps to understand and address these scenarios, particularly in the presence of conflict.

In this regard, the modernization, intensification of extractive processes – including small artisanal fisheries – and convergence of interests, inevitably evolves into tense and conflict. The case of Chile and the Hualaihué Commune, with the presence of Huilliches communities and their inter-action with the salmonid industry, shows how complex even a gradual con-solidation of "modernity" is, particularly in a sociocultural context rooted in values, life expectations and different local development expectation. The tensions among artisanal fishermen, the hydrocarbon industry located in the sea north of Peru, and efforts to create a protected zone also shows evidence of conflictive situations, where the presence of the state and strengths of the deliberative democracy to foresee, mitigate and deactivate conflict are absolutely essential.

The chapters also stress and make a direct or indirect reference to how research in general plays a key role in the effective contribution to problem solving, for example, in the case of migratory species, sea pollution from waste and debris, conflict or industrial fisheries. Existing gaps in information and research on many fronts, limits sound decision-making. Although in Chile, Mexico and Peru, institutions such as the Institute for the Promotion of Fishery, the National Institute for Fishery and Aquaculture and the Insti-tute for the Sea of Peru respectively, play a fundamental role in the genera-tion of data and information, comprehensive research continues to be limited and often focused on specific issues, only relevant for a certain sector.

We finish this publication, returning to a reflection of Professor Castilla in his Prologue, and also highlighted in some of the chapters: the idea that people in Chile, Mexico and Peru are living (or at least have been living) with their backs to the sea. This is changing and with effort and enthusiasm, new and promising scenarios can be perceived in terms of the future for marine and coastal zones. This publication has been aimed towards this goal and we thank those who have read it and made it this far.

Index

Page numbers in **bold** denote tables, those in *italics* denote figures.